Lecture Notes in Computer Science　　9570

Commenced Publication in 1973
Founding and Former Series Editors:
Gerhard Goos, Juris Hartmanis, and Jan van Leeuwen

More information about this series at http://www.springer.com/series/8183

Marina L. Gavrilova · C.J. Kenneth Tan (Eds.)

Transactions on Computational Science XXVII

Springer

Editors-in-Chief

Marina L. Gavrilova
University of Calgary
Calgary, AB
Canada

C.J. Kenneth Tan
Sardina Systems
Tallinn
Estonia

ISSN 0302-9743   ISSN 1611-3349 (electronic)
Lecture Notes in Computer Science
ISBN 978-3-662-50411-6   ISBN 978-3-662-50412-3   (eBook)
DOI 10.1007/978-3-662-50412-3

Library of Congress Control Number: 2015960432

Printed on acid-free paper

This Springer imprint is published by Springer Nature
The registered company is Springer-Verlag GmbH Berlin Heidelberg

LNCS Transactions on Computational Science

Computational science, an emerging and increasingly vital field, is now widely recognized as an integral part of scientific and technical investigations, affecting researchers and practitioners in areas ranging from aerospace and automotive research to biochemistry, electronics, geosciences, mathematics, and physics. Computer systems research and the exploitation of applied research naturally complement each other. The increased complexity of many challenges in computational science demands the use of supercomputing, parallel processing, sophisticated algorithms, and advanced system software and architecture. It is therefore invaluable to have input by systems research experts in applied computational science research.

Transactions on Computational Science focuses on original high-quality research in the realm of computational science in parallel and distributed environments, also encompassing the underlying theoretical foundations and the applications of large-scale computation.

The journal offers practitioners and researchers the opportunity to share computational techniques and solutions in this area, to identify new issues, and to shape future directions for research, and it enables industrial users to apply leading-edge, large-scale, high-performance computational methods.

In addition to addressing various research and application issues, the journal aims to present material that is validated – crucial to the application and advancement of the research conducted in academic and industrial settings. In this spirit, the journal focuses on publications that present results and computational techniques that are verifiable.

Scope

The scope of the journal includes, but is not limited to, the following computational methods and applications:

- Aeronautics and Aerospace
- Astrophysics
- Big Data Analytics
- Bioinformatics
- Biometric Technologies
- Climate and Weather Modeling
- Communication and Data Networks
- Compilers and Operating Systems
- Computer Graphics
- Computational Biology
- Computational Chemistry
- Computational Finance and Econometrics

- Computational Fluid Dynamics
- Computational Geometry
- Computational Number Theory
- Data Representation and Storage
- Data Mining and Data Warehousing
- Information and Online Security
- Grid Computing
- Hardware/Software Co-design
- High-Performance Computing
- Image and Video Processing
- Information Systems
- Information Retrieval
- Modeling and Simulations
- Mobile Computing
- Numerical and Scientific Computing
- Parallel and Distributed Computing
- Robotics and Navigation
- Supercomputing
- System-on-Chip Design and Engineering
- Virtual Reality and Cyberworlds
- Visualization

Editorial

The Transactions on Computational Science journal is part of the Springer journal series *Lecture Notes in Computer Science*, and is devoted to a range of computational science issues, from theoretical aspects to application-dependent studies and the validation of emerging technologies.

The journal focuses on original high-quality research in the realm of computational science in parallel and distributed environments, encompassing the theoretical foundations and the applications of large-scale computations and massive data processing. Practitioners and researchers share computational techniques and solutions in the area, identify new issues, and shape future directions for research, as well as enabling industrial users to apply the techniques presented.

The current volume is devoted to the topic of high-performance computing. It is comprised of eight full papers, covering the specialized domains of cloud middleware, multi-processor systems, quantum computing, optimization, and secure biometric-based encryption methods.

We would like to extend our sincere appreciation to all the reviewers for their work on this regular issue. We especially would like to express our sincere appreciation to Prof. Alexander Bogdanov, TCS Journal Associate Editor, for his facilitation and directional guidance on the issue. Also, our special thanks go to Editorial Assistant Ms. Madeena Sultana, for her dedicated work on collecting papers and communicating with authors. We would also like to thank all of the authors for submitting their papers to the journal and the associate editors and referees for their valuable work.

It is our hope that this collection of eight articles presented in this issue will be a valuable resource for Transactions on Computational Science readers and will stimulate further research into the key area of high-performance computing.

February 2016

Marina L. Gavrilova
C.J. Kenneth Tan

Contents

The Use of Service Desk System to Keep Track of Computational Tasks on Supercomputers

A.V. Bogdanov[1], V.Yu. Gaiduchok[2], I.G. Gankevich[1],
Yu.A. Tipikin[1], and N.V. Yuzhanin[1(✉)]

[1] Saint Petersburg State University, Saint Petersburg 199034, Russia
igankevich@cc.spbu.ru, afk.apmath@gmail.com
[2] Saint Petersburg Electrotechnical University "LETI",
Saint Petersburg 197376, Russia
gvladimiru@gmail.com

Abstract. This paper discusses the unusual use of service desk system–tracking of computational tasks on supercomputers as well as the collection of statistical information on the calculations performed. Particular attention is paid to the possibilities of using such statistics by a supercomputer user as well as the data center staff. The analysis of system requirements for tracking computational tasks and capabilities of service desk and job scheduler systems led to design and implementation of a way to integrate these systems to improve computational tasks management.

1 Introduction

As a result of rapid growth in HPC industry many powerful supercomputers appeared in the last years. They could be very efficient in terms of GFLOPS * hour, but one should always try to keep average resources load at an acceptable level in order to achieve good resource utilization. This implies monitoring, analysis and assessment, and it could be a really challenging task in modern hybrid systems consisting of heterogeneous resources (different CPU architectures, specialized accelerators like NVIDIA Tesla or Intel MIC). In turn multi-user environment imposes load balancing and security problems, logging and accounting tasks.

Problems mentioned above could be partly solved using Portable Batch System (PBS). There are many implementations of PBS that differ from each other in detail, but the main idea is the same. Such systems start jobs in accordance with a scheduler's plan that is based on user preferences, resources load and availability. The scheduler can be a part of entire PBS system or can be a separate software package. It makes a decision about job start time depending on system data collected by PBS. This data updates constantly which is reflected in log files. PBS also store information about jobs: requested resources, resources used, timeframes, etc., and such data is written in the log file too. So, in order to get accounting data one needs to parse large log files and retrieve necessary data.

There is a possibility to send e-mails to users when their jobs are started or stopped (with or without errors). But this is inconvenient too: there are usually too many e-mails

© Springer-Verlag Berlin Heidelberg 2016
M.L. Gavrilova and C.J. Kenneth Tan (Eds.): Trans. on Comput. Sci. XXVII, LNCS 9570, pp. 1–9, 2016.
DOI: 10.1007/978-3-662-50412-3_1

with such information (users usually run many jobs), so this method will simply trash the box. Console PBS commands usually require some Linux skills, so they are inconvenient for inexperienced users too. In this article some approaches for controlling and managing shared HPC resources, which are quite convenient but not traditional, are reviewed.

Similar problems also considered by many authors [1, 2], but they often discuss the HPC in application to the tasks of one research team, so those tasks are analogical and often have a template solution. In our case the information from workload management systems is characterized by its variety and requires more complicated processing methods and tools.

2 Conflict of Interests

HPC center as any social institute involves working with social groups having different goals. Question of retrieving and processing monitoring information is answered differently by the two main groups– users and personnel of HPC center which have different capabilities and different goals. As a result, they need different representations of the monitoring information about the computational jobs. But these groups can also be divided into smaller ones. Users can be divided into executors and leaders while personnel consist of administrators, support engineers and managers. These subgroups have different requirements for the monitoring information too.

For example, user who executes scientific task and directly submits computational jobs to a PBS cluster needs to know which resources are available right now, view job queue, and investigate job failures. He needs to know execution time of finished jobs and amount of consumed resources. Important factor is user-friendly, clear representation of statistics. At the same time such user does not need to know details.

In contrast to executor, research work principal is not interested in particular job. He controls executors, plan the use of the resources which are necessary for work. Last but not the least, he must report to grantor. That is why he needs information about each executor: he needs statistics with the same level of detail as executor and at the same time needs general reports reflecting the overall status of the research work of his scientific group. Tables are more suitable than graphs and diagrams for principal. He usually needs more parameters than executor as he ought to assess the overall process. He can also find automatic reports generation and prediction of resources usage very useful because he has to estimate the efficiency of the work and submit plans for oncoming quarter. He also has to submit requests for additional resources in case of lack of computational power.

On the other side of computational process there are system administrators and managers of the computer center. Administrators should have the possibility to get any information about any job. At the same time they need to get samples from overall statistics about cluster and system usage. For example, it can be necessary for capacity planning and load balancing. Also administrators have to track erroneous jobs and solve any problems related to jobs, system and hardware. So, they need a monitoring system that meets these requirements and also a service desk system which can be used to facilitate work with users and improve work flow. Such system can speed up overall work process and clear interaction with users by creating tickets.

Manager of computational center, in turn, wants to get information about overall resources usage. At the same time, he wants the possibility to assess the resource usage of specific user groups and he does not need detailed reports but an overview of total resource usage during the different time frames. Also, he needs an opportunity to sort reports by each resource and each scientific group.

As one can see, the requirements of mentioned groups have both intersections and differences. Universal system corresponding to the needs of all groups is difficult to implement and the problem is intended to be solved with a system based on a set of services attached to service desk.

3 Computational Infrastructure

Let's take a look at a computational infrastructure of a typical computational center before going deeply into the questions of service desk and monitoring system. Scientific computational clusters are usually equipped with one of Portable Batch System implementations. This system consists of three main parts: PBS Server, PBS Scheduler and PBS MOM with PBS Server being a key component of PBS system. Its task is to receive, modify, create and run jobs. PBS Scheduler decides when and where jobs should be started based on different policies, which administrator can set, and a current cluster load. It interacts with PBS Server in order to get new jobs and with PBS MOMs to retrieve current information about cluster nodes. PBS MOM (or Job Executor) actually starts jobs (creates new processes for jobs on cluster nodes). An instance of PBS MOM runs on each cluster node (moreover, several PBS MOM processes can be started within one node). It is called in this way because such process becomes a parent for all jobs. PBS MOM starts a job when PBS Server requests it. It starts a new session (such a session is identical to a certain degree to a common Linux user session). User needs to install tiny package containing basic PBS client commands. After that jobs can be submitted to a PBS server. All in all, the PBS work can be illustrated in the picture.

Queue is the key component of PBS and all cluster nodes are assigned to a particular queue. Usually there is more than one queue within a PBS system. Nodes of the cluster can be assigned to several queues at the same time. Queues can vary in parameters: priority, time restrictions for jobs, hardware restrictions (e.g. max number of CPUs for a job), etc. Cluster administrator should declare some policy to create queues and assign necessary parameters to them. For example, several queues that have different priority and different time restrictions can be created, i.e. some queues with high priority and severe time restrictions (job can be executed only several hours or minutes), and some queues with low priority and execution time restricted by days or months. So, when a job is submitted, a user can specify the queue based on his preferences. If he wants to compute his task right now, he should choose a queue of the first type.

Server management, submission of jobs and job control are performed using console and all PBS commands can be divided into three groups. The first one is user commands which can be performed by an authorized user. For example, 'qsub' command is used to submit a job to the queue, 'qstat' command is used for retrieving job status. User can also monitor the cluster status using 'pbsnodes' command and some parameters of PBS system.

The main information for customer is his jobs status which shows whether his job was started or not and how much time was consumed. Another set of commands is operator commands. Operator has permissions to monitor and control (hold, alter or terminate) every job. Last but not the least the commands that administrator can execute include all the commands of operator and customer: administrator can control all jobs, perform security tasks and manage cluster nodes.

At this point one can think that PBS is quite enough for performing monitoring of supercomputer systems, however, as with most Linux daemons PBS monitoring capabilities are limited to writing important information to logs. Moreover, in order to get working system that satisfies all user requirements, complies with center policies and is convenient for everyone one should solve some more problems.

One can face many problems while providing users with secure and comfortable access to the cluster. When there are several clusters in the center the problems only grows and expands. How to organize work with clusters, provide scientific data storage and comply with security policies? All these tasks can be done using virtualization. This technology becomes one of the most promising nowadays.

Users of the center are provided with personal virtual machines from which they can submit jobs to a cluster [3]. Also virtual machines can be used not only for job submission, but for data storage, visualization and even for simple pre- and postprocessing of data. Virtual machine can be viewed as an HPC resource when it is migrated to a specialized high-performance SMP node and its resource capacity (RAM, number of CPUs) is expanded accordingly. Such migration neither change configuration of user environment nor it implies explicit movement of data because it is saved in the home directory. So, virtual machine can be seen as a building block of computer center infrastructure.

Described approach has many advantages both for computer center staff and users. Computer center staff benefits from the following capabilities.

- Dynamic resource reallocation.
- Enhanced security.
- VM migration.
- Fault tolerance.
- Flexible configuring.

From the user perspective one can note next advantages.

- Enhanced security.
- High availability.
- Flexible VM configuring.
- Access to customized environment from anywhere.
- Ability to create own virtual clusters.

Data consolidation is one of the today's challenges. Imagine a typical situation for HPC center when many customers have access to many virtual machines and clusters, each user has different home directories on different servers. The other big problem is access to user data from cluster nodes. Both these problems can be solved using single mounted point for user home directory. Separate storage system can be used for this goal. All user's virtual machines mount the home directory from that storage. When user job is started on cluster nodes his home directory is mounted automatically. So, virtual machines and cluster nodes mounts home directory from single place, and user has universal access to his data from all the resources. That is how data consolidation is achieved in our case.

4 Implementation

In its usual setting the service desk provides the single point of ticket reception, ticket workflow and archiving after the solution was found and the ticket was closed [4]. Also the service desk provides the permissions on actions with tickets to the helpdesk staff according their roles, accounts the incident solution time, provides reminders, notifications, ticket escalation mechanism, etc. Service desk system also usually has the knowledge base which contains most frequent incidents and its solutions.

The basis of our research is a model of supercomputer center which provides scientists with computational resources. Scientific work requires detailed report to the grantors, that is why such supercomputer center must provide detailed report and statistics to its users. In such center the service desk system can be the central point of the interaction between supercomputer center staff and users. All correspondence about the incidents also can be processed by service desk.

On the assumption of such position of the service desk it is reasonable to make the website based on the open source CMS as a frontend and the service desk as a backend of the user support system. On our virtual stand OTRS service desk and the website based on the Drupal CMS were integrated using the OTRS GenericInterface and SOAP protocol [5]. To track the computational tasks on the supercomputers the idea of receiving the email reports of PBS as tickets by the service desk was used.

It seems to be simpler to solve the problem without service desk and use direct transfer of the PBS email reports to the CMS database instead. In this way the processing can be done by a custom website module. However, the usage of service desk in this case has strong advantages. On the one hand the database schema development is not needed. On the other hand the service desk system has a convenient "ticket" object with automatic state machine. This object allows us to automatically open a ticket by the keyword in the email header and to close it the same way. And finally the service desk allows the supercomputer center staff to watch the overall picture of the batch system and to operate with the PBS error as with a trouble-ticket.

Also usage of the service desk system as the core of the information system of supercomputer center corresponds the good practices of the ITIL [6]. The integration of service desk with PBS and the website realizes a part of the capacity management.

Basic level of integration is very simple: the PBS server has an installed MTA (mail transfer agent) that sends reports to the service desk from the task executor account. The OTRS service desk has an additional queue and the mail-account connecting it with the email-address. User account on the website has a frame that displays the tickets based on computational tasks in PBS in turn. Basic element of integration between the website and the OTRS is a set of REST web services that get the data from the OTRS database and represents it to the user (Fig. 1).

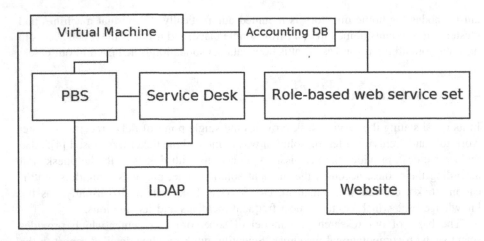

Fig. 1. Integration scheme.

Such configuration works and provides the information, but there are two significant disadvantages: there is no automatic change of ticket states and there are two tickets for every computational task. Large amount of tickets in the dashboard seems uncomfortable for the user. The second disadvantage is inconvenience of the data representation for the user and the lack of statistics and metrics. The first disadvantage can be eliminated by the usage of service desk module for the ticket state automation.

In particular the OTRS SystemMonitoring module can automate ticket state changing. This module gives a possibility to close tickets automatically by the keyword in the header or body of the ticket. The state changing rules are set by the regular expressions and are easy to configure. The purpose of the SystemMonitoring module is the integration of OTRS with different monitoring systems such as Zabbix or Nagios, but it can be used for nonstandard tasks. But this module cannot work with more than one source of information (email account). So it can be used either for the real monitoring or our unusual application, but not for both. That is why usage of this module on the production OTRS server is impossible without the loss of monitoring possibility. Such problem has two appropriate solutions: first is to set up alternative virtual OTRS server only for task tracking using common LDAP server and website interface; the second way to solve the problem is to use the set of postmaster filters based on the regular expressions instead of the SystemMonitoring module. However, the latter approach is complicated with sophisticated regular expression patterns and the resulting system may become unstable or slow due to inefficient processing of the filters.

The first solution allows not only to automate the ticket state changing process but also automatically set the special incident state of the ticket in case of program error keyword in the follow-up or by the timeout. Such incidents can be redirected and then processed by the technical support staff of the computing center. Such possibility is highly important for the system administrators and allows saving the technical support time from the manual batch system monitoring.

Table 1 displays the growth of the efficiency of error processing after the application of our solution. Earlier the period of manual batch system monitoring was approximately 8 h, so the average time of the error detection was 4 h respectively. After deployment the system based on the service desk the manual monitoring is not actual so the average error detection time became 10 min and depends of service desk dashboard refresh time. The average time of the incident processing by the super-computer center staff did not change and still 2 h. So the average error processing time decreased from nearly 6 h to slightly more than 2 h and the helpdesk staff spends their time 3 times more efficiently.

Table 1. Time costs of an error processing.

	Error detection	Incident processing	Total
Before	4 h	2 h	6 h
After	10 min	2 h	2 h 10 min

So we solved the problem of ticket processing automation but the representation of the information is uncomfortable for the user. The simplest way to represent the data is to put it into the tables. But the most comfortable way is to combine tables with graphs and diagrams.

The next problem is the statistics and metrics generation. Relevant information for the user depends on this user role: executor, research work principal, system admin-istrator, computer center manager. Such dependency can be realized by the LDAP technology. LDAP account can contain role information and information about user's priority for the batch system. The OTRS, PBS and either the website uses the common LDAP server to associate the data relative with one user account. For the statistic report creation OTRS Stats module can be used. This tool can be controlled via SOAP, so user can choose reported parameters on the website and generate the report at any time interacting with simple and comfortable GUI.

The question of ordering and recording the activities of various groups of users on the physical computing resources is not a simple question, especially when organizing access to the resource through a heterogeneous system of services. Stats module can be used to gather statistics, but it is not finely tuned and does not allow changing the configuration of the output data "on the fly". For fast and fine-tuning data output, a web service connected to OTRS via the corresponding API can be used.

First, the possible information structure of datacenter can be considered. On the upper level (frontend) there is a portal (website) of center with a functional personal account. The instruments providing control of computational resource consumption will be placed in that account. On the lower (middleware) level there are virtual machines. Through them users execute their intensive computational tasks on the third level of infrastructure– clusters. Such infrastructure allows information about virtual machine resource utilization and information about running and completed tasks on clusters to be obtained and processed. To solve tasks it makes sense to use the ready-made solutions namely PBS for clusters and monitoring modules of VMWare vSphere for VM's.

Personal account is implemented as a separate module of Drupal CMS interacting with databases (CouchDB, ActiveDirectory) through REST web services and direct requests using Sag library. The summary report of work made using datacenter resources contains resource utilization figures, however, datacenter systems that log activity of user virtual machines and applications running on cluster are not linked directly to this account. To display on-demand computing resource usage statistics, a service that would collect all metrics on the user virtual machines and represent this data in the user account in accessible graphical form.

To understand the structure of such implementation data sources for charts should be defined. Suppose that virtual machines accounting is stored separately in the SQL database. It is quite common solution, but this is not the most efficient way of storing such information because of large number of records. Consequently, the request gathering statistics of several years will be executed for an unacceptably long time. A more appropriate way of storing account data is to store the data in a distributed database using staging tables that already contain aggregated data up to current day. Information about tasks executed on a supercomputer is stored as chains of tickets: one for the beginning of the computational task and one for the end. This information is attached to personal account through PHP extensions for CMS and different criteria can be chosen to be displayed. PHP extension then submits request to the web service that contains the data and then pre-generated chart is received as a static picture.

A web service receives a list of user virtual machines from personal account and accounting information from OTRS and a combined chart is created for both cluster and virtual machine utilization. Based on received data web service creates a chart and returns it as a bitmap. The main difficulty lies in the fact that the request has to be completed in a short period of time. That is why the distributed database is a requirement. For chart creation free Java library JfreeChart is used. Such approach allows meeting the needs of executors and research work principal for the visual representation of data in the form of graphs and charts.

Graphical representation gives users more options for an adequate estimation of resources utilization, which further helps to avoid excessive (or lack of) resource allocation, increases efficiency of computational services and prevents from conflicts between customers and computer center staff. As a part of the development process resource quotas can be introduced for each research work group so that virtual machines can be created independently by customers from these groups. In this case, a fine-grained control of resources is transferred to customers' hands.

5   Conclusions

The task of integrating multiple information systems with comfortable user-friendly environment for computational task tracking allows an executor to make necessary quantitative evaluations of his work on the supercomputer. For example such system can be useful to make the report about numerical simulation performed within the confines of a research work or to track the computational resource quotas utilization and make predictions.

Our system will be especially useful for research work principals. Using the system they can evaluate the progress of all computational tasks within the confines of the group, the productivity of every subordinate, the measure and the uniformity of quotas consumption. Also the automated report generator eliminates the need for manual generation of report after the research work ends. Metrics of total resource consumption is included into the report automatically.

Supercomputer center staff also will not be forgotten. Technical support can easily receive information about the total computational resource consumption and make the prognosis or a plan of further services. System keeps the data in the service desk database and allows archiving and making backups to protect the information. Also thanks to the database the reports can cover any time period. Such opportunity is very useful for the computing center managers.

Another advantage of solution based on the service desk system became a possibility to increase the efficiency of solving the program errors on the supercomputers. The time of error recognizing decreases thanks to automatic incident state changing. That is why overall time costs also decreases and the helpdesk staff works more efficient.

Acknowledgements. Research was carried out using computational resources provided by Resource Center "Computer Center of SPbU"(Official web site: http://cc.spbu.ru/.) and supported by Russian Foundation for Basic Research (project N 13-07-00747) and Saint Petersburg State University (projects N 9.38.674.2013, 0.37.155.2014).

References

1. Krompass, S., et al.: Dynamic workload management for very large data warehouses: Juggling feathers and bowling balls. In: Proceedings of the 33rd International Conference on Very Large Data Bases, pp. 1105–1115 (2007)
2. Davison, D., Graefe, G.: Dynamic resource brokering for multi-user query execution. In: Proceedings of the 1995 ACM SIGMOD International Conference on Management of Data, pp. 281–292 (1995)
3. Gayduchok, V.Yu., Bogdanov, A.V., Degtyarev, A.B., Gankevich, I.G., Gayduchok, V.Yu., Zolotarev, V.I.: Virtual workspace as a basis of supercomputer center. In: Proceedings of the 5th Internship Conference "Distributed Computing and Grid-Technologies in Science and Education" (Dubna, 16-21 July, 2012)/ Joint Institute for Nuclear Research (Dubna), pp. 60–66 (2012)
4. Kácha, P.: OTRS: CSIRT WorkFlow Improvements. – CESNET, Technical Report, 10 (2010)
5. Bakker, R., et al.: OTRS 3.3 - Admin Manual (2013). http://ftp.otrs.org/pub/otrs/doc/doc-admin/3.3/en/pdf/otrs_admin_book.pdf. Accessed 7 April 2014
6. Potgieter, B.C., Botha, J.H., Lew, C.: Evidence that use of the ITIL framework is effective. In: 18th Annual Conference of the National Advisory Committee on Computing Qualifications, Tauranga, NZ. C, pp. 160–167 (2005)

Mapping of Subtractor and Adder-Subtractor Circuits on Reversible Quantum Gates

Himanshu Thapliyal[✉]

Department of Electrical and Computer Engineering,
University of Kentucky, Lexington, KY, USA
hthapliyal@uky.edu

Abstract. Reversible arithmetic units such as adders, subtractors and comparators form the essential components of any hardware implementation of quantum algorithms such as Shor's factoring algorithm. Further, the synthesis methods proposed in the existing literature for reversible circuits target combinational and sequential circuits in general and are not suitable for synthesis of reversible arithmetic units. In this paper, we present several design methodologies for reversible subtractor and reversible adder-subtractor circuits, and a framework for synthesizing reversible arithmetic circuits. Three different design methodologies are proposed for the design of reversible ripple borrow subtractor that vary in terms of optimization of metrics such as ancilla inputs, garbage outputs, quantum cost and delay. The first approach follows the traditional ripple carry approach while the other two use the properties that the subtraction operation can be defined as $a - b = \overline{\overline{a} + b}$ and $a - b = a + \overline{b} + 1$, respectively. Next, we derive methodologies adapting the subtractor to also perform addition as selected with a control signal. Finally, a new synthesis framework for automatic generation of reversible arithmetic circuits optimizing the metrics of ancilla inputs, garbage outputs, quantum cost and the delay is presented that integrates the various methodologies described in our work.

1 Introduction

Reversible circuits can generate unique output vector from each input vector, and vice versa, that is, there is a one-to-one mapping between the input and output vectors. One of the major applications of reversible logic lies in quantum computing. A quantum computer will be viewed as a quantum network (or a family of quantum networks) composed of quantum logic gates; each gate performing an elementary unitary operation on one, two or more two-state quantum systems called qubits. Each qubit represents an elementary unit of information; corresponding to the classical bit values 0 and 1. Any unitary operation is reversible and hence quantum networks must be built from reversible logical components [32,51]. Several important metrics need to be considered in the design of reversible circuits the importance of which needs to be discussed. Quantum computers of many qubits are extremely difficult to realize thus the number

© Springer-Verlag Berlin Heidelberg 2016
M.L. Gavrilova and C.J. Kenneth Tan (Eds.): Trans. on Comput. Sci. XXVII, LNCS 9570, pp. 10–34, 2016.
DOI: 10.1007/978-3-662-50412-3_2

of qubits in the quantum circuits needs to be minimized. This sets the major objective of optimizing the number of ancilla input qubits and the number of the garbage outputs in the reversible logic based quantum circuits. The constant input in the reversible quantum circuit is called the ancilla input qubit (ancilla input bit), while the garbage output refers to the output which exists in the circuit just to maintain one-to-one mapping but is not a primary or a useful output. The reversible circuit has other important parameters of quantum cost and delay which need to be optimized.

A synthesis framework in which a single parameter is optimized is inadequate since optimizing one parameter often could be resulting in the degradation of other important parameters. Further, the general reversible synthesis methods proposed in the existing literature target combinational and sequential logic synthesis in general and are not suitable for synthesis of reversible arithmetic units. This is because in arithmetic units such as adders, multipliers, shifters, etc., the choice of the hardware algorithm or the architecture has an impact on the performance and efficiency of the circuit. For instance, in the case of the adder and multiplier designs, the choice of the scheme such as carry look-ahead adder, carry skip adder, array multiplier, wallace tree multiplier etc. becomes critical and in reversible circuit design, the optimization of multiple parameters is impacted by the scheme. Thus, there is a need for research towards developing new design and synthesis methods for realization of reversible arithmetic circuits and a synthesis framework in which multiple parameters can be optimized. In reversible logic based circuit design, parameters such as ancilla inputs, garbage outputs, quantum cost and delay are important which are completely different from the traditional parameters of speed, power and chip area used in conventional computing.

Reversible arithmetic units such as adders, subtractors, multipliers form the essential component of a quantum computing system. Researchers have addressed the design of reversible adders, multipliers, sequential circuits such as in [3,4,15,16,45–47]. In this paper, we present several design methodologies for reversible subtractor and reversible adder-subtractor circuits, and a framework for synthesizing reversible arithmetic circuits. Three different design methodologies are proposed for the design of reversible ripple borrow subtractor that vary in terms of optimization of metrics such as ancilla inputs, garbage outputs, quantum cost and delay. The first approach follows the traditional ripple carry approach while the other two use the properties that the subtraction operation can be defined as $a - b = \overline{\overline{a} + b}$ and $a - b = a + \overline{b} + 1$, respectively. Next, we derive methodologies adapting the subtractor to also perform addition as selected with a control signal. Novel designs of reversible half subtractor, 1 bit reversible full subtractor and 1 bit reversible full adder are also presented optimizing the metrics of ancilla inputs, garbage outputs, quantum cost and the delay. Finally, a new synthesis framework for automatic generation of reversible arithmetic circuits optimizing the metrics of ancilla inputs, garbage outputs, quantum cost and the delay is presented that integrates the various methodologies described in our work. To the best of our knowledge this is the first attempt towards the proposal of a framework for the synthesis of reversible arithmetic circuits.

The paper is organized as follows: Sect. 2 presents the basic reversible gates, quantum cost, delay and the prior works. Section 3 presents the first approach that discusses the design of reversible subtractor using conventional ripple borrow approach. Section 4 presents the second approach that discusses the design of reversible subtractor circuit based on reversible adder without input carry. Section 5 presents the third approach that discusses the design of reversible subtractor circuit based on reversible adder with input carry. Section 6 shows the comparison of n bit reversible subtractors. The design of unified reversible adder-subtractor using the three approaches is discussed in Sect. 7. Section 8 presents the integration of proposed design methodologies as a synthesis framework while the discussion and conclusions are provided in Sect. 9.

2 Basics

There are many popular 3 inputs and 3 outputs reversible gates such as the Fredkin gate [11], the Toffoli gate [49] and the Peres gate [34], etc. Any 3 inputs and 3 outputs reversible gate can be realized using 1×1 NOT gate, and 2×2 reversible gates such as Controlled-V and Controlled-V^+ (V is a square-root-of NOT gate and V^+ is its hermitian) and the Feynman gate which is also known as the Controlled NOT gate (CNOT). Quantum cost is the one of the metrics that is used compare different 3×3 reversible logic gates. The quantum cost of all reversible 1×1 and 2×2 gates is taken as unity [17,24,40], while the quantum cost of a 3×3 reversible gate can be calculated by counting the numbers of NOT, Controlled-V, Controlled-V^+ and CNOT gates required in its implementation.

2.1 The NOT Gate

A NOT gate is a 1×1 gate represented as shown in Fig. 1(a). Since it is a 1×1 gate, its quantum cost is unity.

2.2 The Controlled-V and Controlled-V^+ Gates

The controlled-V gate is shown in Fig. 1(b). In the controlled-V gate, when the control signal A = 0 then the qubit B will pass through the controlled part unchanged, i.e., we will have Q = B. When A = 1 then the unitary operation $V = \frac{i+1}{2} \left(\begin{smallmatrix} 1 & -i \\ -i & 1 \end{smallmatrix} \right)$ is applied to the input B, i.e., Q = V(B). The controlled-V^+ gate is shown in Fig. 1(c). In the controlled-V^+ gate when the control signal A = 0 then the qubit B will pass through the controlled part unchanged, i.e., we will have Q = B. When A = 1 then the unitary operation $V^+ = V^{-1}$ is applied to the input B, i.e., Q = V^+(B).

The V and V^+ quantum gates have the following properties:

$$V \times V = NOT$$
$$V \times V^+ = V^+ \times V = I$$
$$V^+ \times V^+ = NOT$$

The properties above show that when two V gates are in series they will behave as a NOT gate. Similarly, two V^+ gates in series also function as a NOT gate. A V gate in series with V^+ gate, and vice versa, is an identity. For more details of the V and V^+ gates, the reader is referred to [17,32].

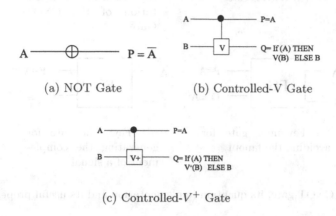

(a) NOT Gate   (b) Controlled-V Gate

(c) Controlled-V^+ Gate

Fig. 1. The NOT and the Controlled-V and Controlled-V^+ gates

2.3   The Feynman Gate (CNOT Gate)

The Feynman gate (FG) or the Controlled-NOT gate (CNOT) is a 2 inputs 2 outputs reversible gate having the mapping (A, B) to (P = A, Q = A ⊕ B) where A, B are the inputs and P, Q are the outputs, respectively. Since it is a 2 × 2 gate, it has a quantum cost of 1. Figure 2(a) and (b) shows the block diagrams and quantum representation of the Feynman gate. The Feynman gate can be used for copying the signal thus avoiding the fanout problem in reversible logic as shown in Fig. 2(c). Further, it can be also be used for generating the complement of a signal as shown in Fig. 2(d).

2.4   The Toffoli Gate

The Toffoli Gate (TG) is a 3 × 3 two-through reversible gate as shown in Fig. 3(a). Two-through means two of its outputs are the same as the inputs with the mapping (A, B, C) to (P = A, Q = B, R = A·B ⊕ C), where A, B, C are inputs and P, Q, R are outputs, respectively. The Toffoli gate is one of the most popular reversible gates and has the quantum cost of 5 as shown in Fig. 3(c) [49]. The quantum cost of Toffoli gate is 5 as it needs 2 V gates, 1 V^+ gate and 2 CNOT gates to implement it. The graphical notation of the Peres gate is shown in Fig. 3(b).

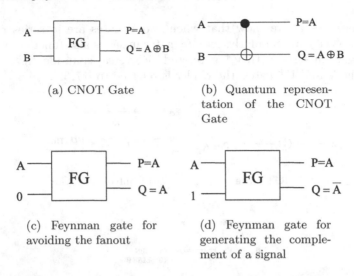

(a) CNOT Gate

(b) Quantum represen-
tation of the CNOT
Gate

(c) Feynman gate for
avoiding the fanout

(d) Feynman gate for
generating the comple-
ment of a signal

Fig. 2. CNOT gate, its quantum implementation and its useful properties

2.5   The Peres Gate

The Peres gate is a 3 inputs 3 outputs (3×3) reversible gate having the mapping
(A, B, C) to ($P = A$, $Q = A \oplus B$, $R = A \cdot B \oplus C$), where A, B, C are the inputs
and P, Q, R are the outputs, respectively [34]. Figure 4(a) shows the Peres gate
and Fig. 4(c) shows the quantum implementation of the Peres gate (PG) with
quantum cost of 4 [17]. The quantum cost of Peres gate is 4 since it requires 2
V^+ gates, 1 V gate and 1 CNOT gate in its design. In the existing literature,
among the 3×3 reversible gates, the Peres gate has the minimum quantum cost.
The graphical notation of the Peres gate is shown in Fig. 4(b).

2.6   Delay Computation in Reversible Logic Circuits

Delay is another important parameter that can indicate the efficiency of
reversible circuits. Here, delay represents the critical delay of the circuit. In
many of the earlier works on reversible combinational circuits such as in [3,19],
the delays of each reversible gate such as 2×2, 3×3 and 4×4 reversible gates,
all are considered to be of unit delay irrespective of their computational complexity. This is not fair for comparison as delay will vary according to the complexity
of a reversible gate. In our delay calculations, we use the logical depth as the
measure of the delay [28]. The delays of all 1×1 gate and 2×2 reversible gate
are taken as unit delay called Δ. Any 3×3 reversible gate can be designed from
1×1 reversible gates and 2×2 reversible gates, such as the CNOT gate, the
Controlled-V and the Controlled-V^+ gates. Thus the delay of a 3×3 reversible
gate can be computed by calculating its logical depth when it is designed from
smaller 1×1 and 2×2 reversible gates. Figure 3(c) shows the logic depth in the

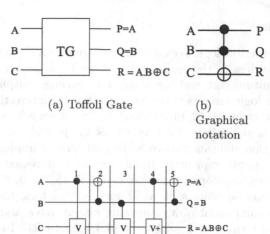

(a) Toffoli Gate

(b) Graphical notation

(c) Quantum implementation

Fig. 3. The Toffoli gate and its quantum implementation

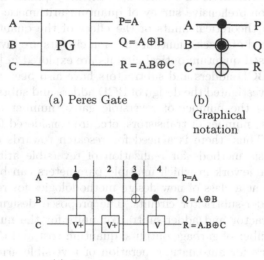

(a) Peres Gate

(b) Graphical notation

(c) Quantum implementation

Fig. 4. The Peres gate and its quantum implementation

quantum implementation of Toffoli gate. Thus, it can be seen that the Toffoli gate has the delay of 5 Δ. Each 2×2 reversible gate in the logic depth contributes to 1 Δ delay. Similarly, Peres gate shown in Fig. 4(c) has the logic depth of 4 that results in its delay as 4 Δ.

2.7   Prior Works

The research on reversible logic is expanding towards both design and synthesis. Researchers have addressed the optimization of reversible logic circuits from the perspective of quantum cost and the number of garbage outputs. In the synthesis of reversible logic circuits there has been several interesting attempts in the literature such as in [10,13,14,22,25,35,37,38]. Researchers have also proposed new synthesis algorithms that synthesizes an optimal circuit for any 4-bit reversible specification and can also synthesizes all optimal implementations [12]. The designs of reversible sequential circuits are also addressed in literature in which various latches, flip-flops, etc. are designed [7,15,36,39,45].

Reversible arithmetic units such as adders, subtractors, multipliers which form the essential component of a computing system have also been designed in binary as well as ternary logic such as in [3,4,9,16,19–21]. In [8], researchers have designed the quantum ripple carry adder having no input carry with one ancilla input bit. In [42,43], the researchers have investigated new designs of the quantum ripple carry adder with no ancilla input bit and improved delay. In [50], the measurement based design of carry look-ahead adder is presented while in [27] the concept of arithmetic on a distributed-memory quantum multicomputer is introduced. A comprehensive survey of quantum arithmetic circuits can be found in [41]. The theoretical limits of the effect of the quantum interaction distance on the speed of exact quantum addition circuits are investigated in [6]. Reversible logic based quantum optical circuits are explored in [52].

The design of BCD adders and subtractors have also been attempted. The researchers have investigated the design of BCD adders and subtractors in which parameters such as the number of reversible gates, number of garbage outputs, quantum cost, number of transistors, etc. are considered for optimization [2,3,18,29,30,48]. Thus, there is a need for research towards developing new design and synthesis methods for realization of reversible arithmetic circuits and a synthesis framework in which multiple parameters can be optimized. In this work, we present a class of new design methodologies for reversible binary subtractor and adder-subtractor circuits. The proposed design methodologies optimize the subtractor and adder-subtractor units for the number of ancilla inputs and the number of garbage outputs, quantum cost and the delay. A new synthesis framework for automatic generation of reversible arithmetic circuits optimizing the metrics of ancilla inputs, garbage outputs, quantum cost and the delay based on the user specifications is also presented.

3   Approach 1: Design of Reversible Subtractor Using Conventional Ripple Borrow Approach

Before discussing the existing design of reversible half subtractor, the basic working of a half subtractor is illustrated. Let A and B are two binary numbers. The half subtractor performs A-B operation. Table 1 shows the truth table of the half subtractor. The output of the XOR gate produces the difference between

A and B. The output of the AND gate $\bar{A} \cdot B$ produces a Borrow. Thus, the output function will be $Borr = \bar{A} \cdot B$; $Diff = A \oplus B$. In the existing literature, the reversible half subtractor as shown in Fig. 5 is designed from 2 CNOT gates (2 Feynman gates) and 1 Toffoli gate [31]. The design in [31] is the most widely used design of quantum half subtractor [1, 33]. The existing design of the reversible half subtractor in [31] has the quantum cost of 7 and delay of 7 Δ, while the existing design in [44] has the quantum cost of 6 and delay of 6 Δ. In this work, we propose the reversible half subtractor design based on a new quantum implementation of the reversible TR gate. The reversible TR gate is a 3 inputs 3 outputs gate having inputs to outputs mapping as (P = A, Q = $A \oplus B$, $R = A \cdot \bar{B} \oplus C$) as shown in Fig. 6(a). The implementation of the TR gate with 2×2 reversible gates is shown in Fig. 6(c) which shows that the proposed TR gate has quantum cost of 4 and delay of 4 Δ. It is to be noted that the upper bound on the quantum cost of the TR gate was estimated as 6 in [44]. The graphical notation of the TR gate is shown in Fig. 6(b).

Table 1. Truth table of half subtractor

A	B	Borr	Diff
0	0	0	0
0	1	1	1
1	0	0	1
1	1	0	0

Figure 7(a) shows the working of the TR gate as a reversible half subtractor. As shown in Fig. 7(b), the TR gate implements the reversible half subtractor with quantum cost of 4, delay of 4 Δ and 0 garbage outputs (the inputs regenerated at the outputs are not considered as garbage outputs). A comparison of the reversible half subtractors is shown in Table 3. Thus proposed design achieves 43 % reduction in terms of quantum cost (QC) and delay compared to design presented in [31], while the improvement compared to design presented in [44] is 33 % both in terms of the quantum cost (QC) and the delay (Table 3).

3.1   Design of Reversible Full Subtractor

To subtract three binary numbers, one can use a full subtractor which realizes the operation Y = A-B-C. The truth table of the full subtractor is shown in Table 4. This gives the equation of the borrow and difference as follows: $Diff = A \oplus B \oplus C$; $Borr = A \cdot \bar{B} \oplus \overline{A \oplus B} \cdot C$. In the existing literature, the reversible full subtractor is designed with 2 Toffoli gates, 3 Feynman gates and 2 NOT gates [5]. The existing design of reversible full subtractor is shown in Fig. 8. Thus, the existing reversible full subtractor has the quantum cost of 15, delay of 15 Δ.

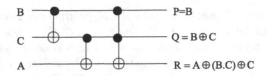

Fig. 5. Existing design for quantum half subtractor [31]

Table 2. Truth table for the TR gate

A	B	C	P	Q	R
0	0	0	0	0	0
0	0	1	0	0	1
0	1	0	0	1	0
0	1	1	0	1	1
1	0	0	1	1	1
1	0	1	1	1	0
1	1	0	1	0	0
1	1	1	1	0	1

(a) TR Gate   (b) Graphical notation

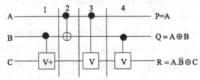

(c) Quantum implementation

Fig. 6. The TR Gate and its quantum implementation

In this work, we propose the design of the reversible full subtractor in Fig. 9. It requires two TR gates to design a reversible full subtractor with no garbage output and 1 ancilla input. The quantum realization of the TR gate based reversible full subtractor is shown in Fig. 10(a). From Fig. 10(a), we can see that the TR gate based reversible full subtractor has the quantum cost of 8 with delay of 8 Δ. As can be seen in the Fig. 10(a) the fourth gate (V gate) and the fifth gate (V^+ gate) are in series thus forming an identity and can be removed.

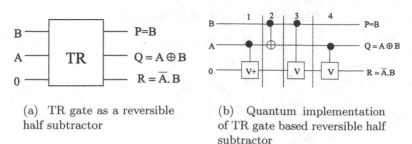

(a) TR gate as a reversible half subtractor

(b) Quantum implementation of TR gate based reversible half subtractor

Fig. 7. Improved design of the TR gate based reversible half subtractor

Table 3. A comparison of reversible half subtractors

	QC	Delay
Design proposed in [31]	7	7
Design proposed in [44]	6	6
Proposed design	4	4
Improvement in % w.r.t [31]	43	43
Improvement in % w.r.t [44]	33	33

This results in a new optimized design of TR gate based reversible full subtractor with quantum cost of 6 and delay of 6 Δ as shown in Fig. 10(b). Further, in the optimized design shown in Fig. 10(b) few of the gates can be moved by applying the move rules discussed in [23] without affecting the functionality of the circuit. This results in two pairs of V and CNOT gates operating in parallel, and generate a new design of the reversible full subtractor as shown in Fig. 10(c) that has the delay of 4 Δ. Thus, compared to the existing design [5], the proposed reversible full subtractor design based on TR gate has an improvement ratios of 60 % and 73.33 % in terms of quantum cost(QC) and delay, respectively. The improvements compared to existing design [44] are 50 % and 66.67 % in terms of the quantum cost and the delay. All the existing designs of the reversible full subtractor also have no garbage outputs and 1 ancilla input. The results are summarized in Table 5.

3.2   Design of N Bit Reversible Subtractor

Once we have designed the 1 digit reversible full subtractor, the n bit reversible subtractor can be designed by cascading a reversible half subtractor followed by a reversible full subtractor, followed by a reversible full subtractor, and so on using the ripple borrow approach. A n bit reversible subtractor subtracting n bit numbers a_i and b_i where $0 \leq i \leq n - 1$ is shown in Fig. 11. The design of n bit reversible subtractor using the ripple borrow approach has $n - 1$ ancilla inputs, $n - 1$ garbage outputs, quantum cost of $6n - 2$ and delay of $4n$ Δ. Table 6 illustrates that the proposed design is better than the existing design in terms of the quantum cost and the delay.

Table 4. Truth table of full subtractor

A	B	C	Borr	Diff
0	0	0	0	0
0	0	1	1	1
0	1	0	1	1
0	1	1	1	0
1	0	0	0	1
1	0	1	0	0
1	1	0	0	0
1	1	1	1	1

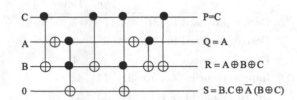

Fig. 8. Existing design of quantum full subtractor [5]

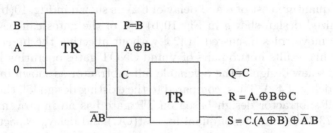

Fig. 9. The TR gate as a full subtractor

Table 5. A comparison of reversible full subtractors

	QC	Delay
Design proposed in [5]	15	15
Design proposed in [44]	12	12
Proposed Design	6	4
Improvement in % w.r.t [5]	60	73.33
Improvement in % w.r.t [44]	50	66.67

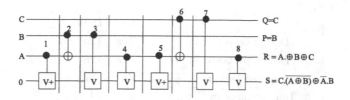

(a) Quantum implementation of TR gate based reversible full subtractor

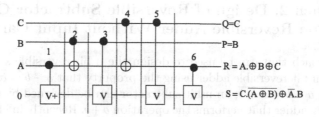

(b) Optimized Quantum implementation of TR gate based reversible full subtractor

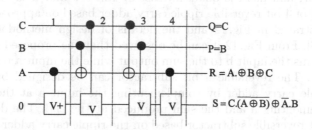

(c) Delay optimization of TR gate based reversible full subtractor

Fig. 10. Quantum implementation of TR gate based reversible full subtractor

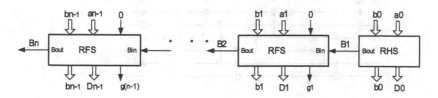

Fig. 11. n bit reversible full subtractor

Table 6. A comparison of n bit reversible full subtractors

	QC	Delay
Design proposed in [5]	15n	15n
Design proposed in [44]	12n	12n
Proposed Design	6n-2	4n
Improvement in % w.r.t [5]	60	73.33
Improvement in % w.r.t [44]	50	66.67

4   Approach 2: Design of Reversible Subtractor Circuit Based on Reversible Adder Without Input Carry

Another approach that can be used to design the n bit reversible subtractor is based on the n bit reversible adder using the property that $a - b = \overline{\overline{a} + b}$. Using this approach $a - b = \overline{\overline{a} + b}$, an n bit subtractor can be designed by using the n bit ripple carry adder that performs the operation $a+b$. Recently in [47], we have proposed the efficient design of the n bit ripple carry adder that performs the addition of two n bit numbers a and b. For n bit addition, the approach presented in [47] designs the ripple carry adder circuit without any ancilla input and the garbage output, and has the quantum cost of $13n - 8$ and delay of $(11n - 9)$ Δ. The design of 4 bit reversible ripple carry adder based on approach proposed in [47] is illustrated in Fig. 12 and the details of design methodology can be referred in [47]. From Fig. 12 it can be observed that the proposed ripple carry adder transforms the input b to the sum output while the input a is regenerated at the output. The reversible n bit subtractor can be designed based on the proposed ripple carry adder by complementing the input a at the start, and finally complementing a and the sum produced at the end. The design of the proposed n bit reversible subtractor based on the ripple carry adder proposed in [47] is illustrated in Fig. 13. Thus the proposed n bit reversible subtractor has the quantum cost of $13n - 8+3n = 16n - 8$, while the delay is $11n - 9+n+n = (13n - 9)$ Δ.

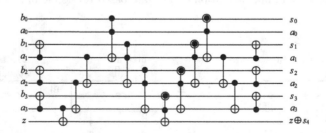

Fig. 12. Proposed reversible 4 bit adder without input carry

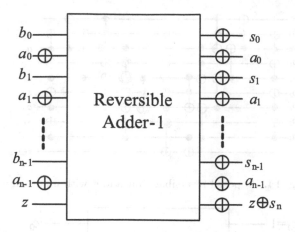

Fig. 13. Proposed reversible n bit subtractor based on approach 2

5   Approach 3: Design of Reversible Subtractor Circuit Based on Reversible Adder with Input Carry

Another possible approach to design an n bit subtractor that can subtract two n bit numbers a and b is to use the property $a - b = a + \bar{b} + 1$. Thus, an n bit subtractor can be designed by using the n bit ripple carry adder with input carry that performs the operation $a + b + c_0$ where a and b are n bit numbers and c_0 is the input carry hardwired as $c_0 = 1$. Recently in [47], we have proposed an efficient design of the n bit ripple carry adder with input carry that performs the addition of two n bit numbers a and b and input carry c_0. For n bit addition, the approach presented in [47] designs the n bit ripple carry adder with input carry with quantum cost of $15n - 6$, while the propagation delay of the design is $(9n + 1)\Delta$. The design of 4 bit reversible ripple carry adder based on approach proposed in [47] is illustrated in Fig. 14. The details of design methodology can be referred in [47]. From Fig. 14 it can be observed that the proposed ripple carry adder transforms the input b to the sum output while the input a and carry input c_0 is regenerated at the outputs. The reversible n bit subtractor circuit based on the ripple carry adder with input carry can be designed by complementing the input b at the start and hardwiring the input $c_0 = 1$. The design of the proposed n bit reversible subtractor based on the ripple carry adder proposed in [47] is illustrated in Fig. 15. Thus the proposed n bit reversible subtractor has the quantum cost of $15n - 6 + n = 16n - 6$, while the delay is $(9n + 1) + 1 = (9n + 2)\,\Delta$.

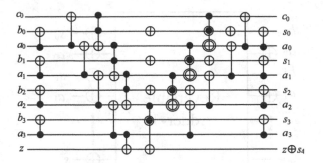

Fig. 14. Proposed reversible 4 bit adder with input carry

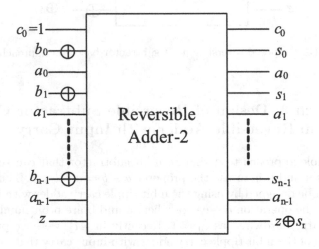

Fig. 15. Proposed reversible n bit subtractor based on approach 2

6   Comparison of n Bit Reversible Subtractor

As illustrated above the n bit reversible subtractor based on approach 1 has the quantum cost of $6n - 2$, delay of $4n\ \Delta$, n-1 ancilla inputs and $n - 1$ garbage outputs. The design of n bit reversible subtractor based on approach 2 has the quantum cost of $16n - 8$, delay of $13n - 9\ \Delta$, 0 ancilla inputs and 0 garbage outputs. The design of n bit reversible subtractor based on approach 3 has the quantum cost of $16n - 6$, delay of $9n + 2\ \Delta$, 1 ancilla input and 1 garbage output. Further, in the existing literature there are designs of reversible n bit subtractor which are based on the approach similar to approach 2. The n bit reversible subtractor proposed in [8] has 1 ancilla input, 1 garbage output, quantum cost of $20n - 12$ and delay of $12n\ \Delta$. The n bit reversible subtractor proposed in [42] has the quantum cost of $29n - 29$ and delay of $(26n-27)\ \Delta$ and is designed without any ancilla input and garbage output. The reversible subtractor based on adder presented in [43] has the quantum cost of $18n - 9$ and delay of $(15n - 7)\ \Delta$ and is also designed without any ancilla input and the garbage output.

Table 7. A comparison of n bit reversible ripple borrow subtractors

	Ancilla input	Garbage outputs	Quantum cost	Delay Δ
This work 1	n-1	n-1	6n-3	6n-3
This work 2	0	0	16n-8	13n-9
This work 3	1	1	16n-6	9n+2
Cuccaro et al. 2004 [8]	1	1	20n-12	12n
Takahashi and Kunihiro 2005 [42]	0	0	29n-29	26n-27
Takahashi et al. 2009 [43]	0	0	18n-9	15n-7

1 represents proposed design of n bit reversible subtractor based on approach 1
2 represents proposed design of n bit reversible subtractor based on approach 2
3 represents proposed design of n bit reversible subtractor based on approach 3
1 is the most efficient design in terms of quantum cost and the delay but has overhead
   in terms of ancilla inputs and the garbage outputs
2 has only 1 ancilla input and 1 garbage output and is better than its existing counterpart in [8] (which is also designed with 1 ancilla input and 1 garbage output) in
   terms of quantum cost and the delay
3 has only 0 ancilla input and 0 garbage output and is better than its existing counterparts in [42,43] (which are also designed with 0 ancilla input and 0 garbage output)
   in terms of quantum cost and the delay

Table 7 summarizes the comparison of all reversible n bit subtractor. The comparison shows the proposed designs of the n bit reversible subtractor excels
the existing design of reversible subtractor in various parameters. The proposed
design based on approach 1 is most efficient in terms of delay and quantum cost
while sacrificing the ancilla inputs and the garbage outputs. The proposed design
based on approach 3 having 1 ancilla input and 1 garbage output is better than
its existing counterpart in [8] in terms of quantum cost and delay. The proposed
design 2 based on approach 2 without any ancilla inputs and the garbage outputs
is better than its existing counterparts proposed in [42,43] in terms of quantum
cost and delay.

7   Design of Unified Reversible Adder-Subtractor

In this section, we discuss the design of unified reversible adder-subtractor that
can work as an adder as well as a subtractor depending on the value of the
control signal (*ctrl*).

7.1   Design of Reversible Adder-Subtractor Based on Approach 1

Before presenting the design of an n bit reversible adder/subtractor we propose
a new design of the 1 bit reversible full adder that forms the basic building block
in the complete design of the adder-subtractor. In the existing literature the
design of 1 bit full adder presented in [23] is one of the most efficient design in
terms of delay. As shown in Fig. 16(a), the design presented in [23] has the sum

and carry outputs produced at R and S outputs, respectively. The reversible full adder in [23] has delay of 4 Δ, quantum cost of 6, needs 1 ancilla input and 1 garbage output (g1). The output Q that produces the function $Q = A \oplus B$ is considered as the garbage output. Even though the design in [23] is efficient in terms of delay, quantum cost and the ancilla input, it has the drawback of producing a garbage output. Thus, in this work we present a new efficient design of a 1 bit reversible full adder with quantum cost of 6, delay of 4 Δ, 1 ancilla input and without any garbage output as illustrated in Fig. 16(b). Thus, the proposed design of the 1 bit reversible full adder is better than the existing design presented in [23]. The output P = C is not considered as the garbage output because in the full adder inputs A,B and C can be exchanged with each other without affecting the functionality of the design. Thus in Fig. 16(b) by passing input B in place of input C, the output P will have value P = B and will not be considered as a garbage output (the regenerated inputs are not considered as the garbage outputs [11]). Once we have an efficient design of 1 bit reversible full adder (RFA), an n bit unified adder-subtractor can be designed as illustrated in Fig. 17. The design uses n 1 bit reversible carry adder (RFA) cascaded as a chain and functions as an adder or a subtractor depending on the values passed by n CNOT gates controlled by the signal labelled as *ctrl*. When the signal *ctrl* = *0* the design works as an n bit reversible adder because the CNOT gates pass the input b in its normal form. When *ctrl = 1*, the design works as an n bit reversible

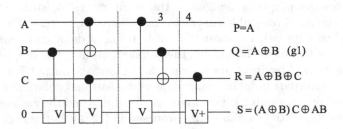

(a) Design of 1 bit reversible full adder proposed in [23] with 1 garbage output

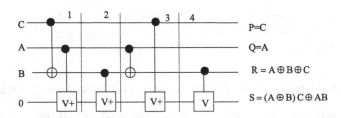

(b) Proposed design of 1 bit reversible full adder without any garbage output

Fig. 16. Designs of 1 bit reversible full adder

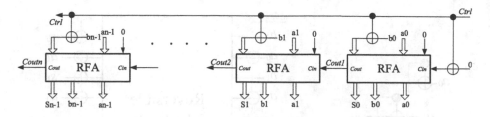

Fig. 17. Proposed design of n bit reversible adder-subtractor based on approach 1

subtractor because the value of the input b is complemented by the CNOT gates thus performing the operation $a - b = a + \bar{b} + 1$. The proposed design of unified reversible adder-subtractor based on approach 1 has the quantum cost of $7n + 1$ and delay of $(4n + 1)$ Δ, while there are $n + 1$ ancilla inputs and no garbage outputs.

7.2   Design of Reversible Adder-Subtractor Based on Approach 2

As discussed above, the approach 2 of the design of an n bit reversible subtractor uses the property $a - b = \bar{a} + b$. The design of n bit reversible subtractor based on approach 2 is shown earlier in Fig. 13. The design of the n bit reversible full subtractor based on approach 2 can be converted to an n bit adder-subtractor by using a control signal that controls the complementing of the input a at the start, and also the complementing of the input a and the output sum produced at the end. An example of this strategy that shows how CNOT gates can be used for controlling the signals is illustrated in Fig. 18(a). Figure 18(a) illustrates that if $ctrl = 1$ the design will complement the controlled input a otherwise the controlled input is passed as such. Using this strategy of using the control signal $ctrl$ to define the add or subtract operation, the proposed design of an n bit reversible adder-subtractor is illustrated in Fig. 18(b). As the CNOT gate has the quantum cost of 1 and delay of 1 Δ, the proposed design of n bit reversible adder-subtractor has quantum cost and delay as same as reversible n bit subtractor based on approach 2. The quantum cost of n bit reversible adder-subtractor based on approach 2 is $13n - 8 + 3n = 16n - 8$ while the delay is $11n - 9 + n + n = (13n - 9)$ Δ.

7.3   Design of Reversible Adder-Subtractor Based on Approach 3

The approach 3 designs an n bit subtractor using the property $a - b = a + \bar{b} + 1$. An n bit subtractor designed by using the n bit ripple carry adder with input carry that performs the operation $a + b + c_0$ where a and b are n bit number and c_0 is the input carry hardwired as $c_0 = 1$ is illustrated earlier in Fig. 15. The design of the n bit reversible full subtractor based on approach 3 can be converted to an n bit adder-subtractor by using a control signal $ctrl$ that controls the complementing of the input b at the start and can also add 1 to the adder to perform the n bit subtraction. The design of n bit adder-subtractor based

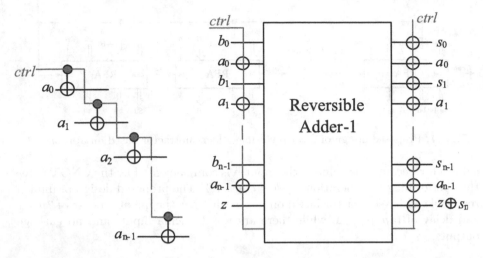

(a) CNOT gates based controlling of the signals

(b) Proposed design of n bit reversible adder-subtractor based on approach 1

Fig. 18. Design of 1 bit reversible full adder and n bit reversible adder-subtractor

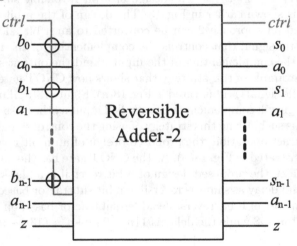

Fig. 19. Proposed reversible n bit adder-subtractor based on Approach 3

on approach 3 is illustrated in Fig. 17. As illustrated in Fig. 17 if $ctrl = 1$ the design will complement the controlled input b and add 1 and will work as n bit reversible subtractor. Otherwise when $ctrl = 0$ the controlled input b is passed as such and the design will work as n bit reversible full adder. The proposed n bit reversible adder-subtractor has the quantum cost and delay as same as n bit reversible subtractor based on approach 3 and is designed without any ancilla input and the garbage output (Fig. 19).

7.4 Comparison of N Bit Reversible Adder-Subtractor

As discussed above the n bit reversible adder-subtractor based on approach 1 has the quantum cost of 7n+1, delay of $(4n+1)$ Δ, n+1 ancilla inputs and 0 garbage outputs. The design of n bit reversible adder-subtractor based on approach 2 has the quantum cost of $16n - 8$, delay of $(13n - 9)$ Δ, 0 ancilla inputs and 0 garbage outputs. The design of n bit reversible adder-subtractor based on approach 3 has the quantum cost of $16n - 6$, delay of $(9n + 2)$ Δ, 0 ancilla inputs and 0 garbage outputs. Thus, design based on approach 1 is most efficient in terms of delay. The results shows that the design based on approach 3 is most efficient as it has 0 ancilla inputs and 0 garbage outputs and has less quantum cost and delay compared to the design based on approach 2. The results shows that the approach 3 will provide an efficient way of designing an n bit reversible adder-subtractor. All the results are summarized in Table 8.

Table 8. A comparison of n bit reversible adder-subtractors

	Ancilla input	Garbage outputs	Quantum cost	Delay Δ
This work 1	n+1	0	7n+1	4n+1
This work 2	0	0	16n-8	13n-9
This work 3	0	0	16n-6	9n+2

1 represents proposed design of n bit reversible adder-subtractor based on approach 1

2 represents proposed design of n bit reversible adder-subtractor based on approach 2

3 represents proposed design of n bit reversible adder-subtractor based on approach 3

8 Integration of Proposed Design Methodologies as a Synthesis Framework

The three different design approaches of the reversible subtractor and reversible adder-subtractor illustrates that a set of design methodologies corresponding to the various arithmetic and logic units can be developed optimizing the different metrics. In this paper, we propose to integrate them as well as the various design methodologies for arithmetic and logic units as a software tool suite. The synthesis portion consists of the algorithms that embed the proposed design methodologies and do not follow the traditional approach due to the variability in the various design methodologies covering a wide range of arithmetic and logic circuits. The proposed synthesis of reversible arithmetic circuits will not be unified and based on a specific synthesis technique, rather will be a collection of algorithms based on the developed design methodologies which is somewhat similar to the approach discussed in [26]. The proposed synthesis framework is discussed below which will evolve and adapt in future based on the needs of the researchers working in the area of reversible arithmetic circuits. The proposed synthesis framework will also consist of design methodologies of reversible arithmetic units proposed by other researchers working in the area of reversible arithmetic circuits.

The framework illustrated in Fig. 20 will consist of design methodologies for (1) reversible adder circuits; (2) reversible multiplier circuits; (3) reversible modular and exponentiation circuits; (4) reversible barrel shifter and comparator circuits; (5) reversible floating point units; (6) other important reversible arithmetic circuits. *The framework currently supports the synthesis of reversible adders, reversible subtractors, reversible adder-subtractors based on the methodologies proposed in this work and also the synthesis of reversible binary and BCD adder circuits proposed in* [8, 42, 43, 47].

The proposed framework consists of three main components: synthesis engine, code generation and the simulation engine.

8.1   Synthesis Engine

The synthesis engine has the following steps: (1) The designer specifies the design requirement such as the parameters to optimize (number of ancilla inputs, garbage outputs, quantum cost, delay, etc.) along with the design scheme. For example, the designer can specify the requirement as 64 bit adder based on carry look-ahead scheme; (2) Based on user specification, with a top level python script, the system searches a look-up table to calculate the cost of the design methodology for the particular scheme of the reversible circuit in terms of the number of ancilla inputs, garbage outputs, quantum cost and delay. The look up table has the generalized cost functions for all the designs and design parameters. If the designer requirement is met, the framework proceeds to the next step, or else another scheme or architecture can be explored. For example, if the carry look-ahead scheme does not meet the requirements the user can explore carry skip or another scheme; (3) We have created a library of all the reversible gates coded in Verilog HDL such as the Fredkin gate, the Toffoli gate, the Peres gate, the TR gate, the Feynman gate etc. If a new reversible gate is proposed in the literature, it can be easily added to the library. In the integrated framework, a built-in library of the Verilog codes of the various design methodologies of different reversible arithmetic circuits is created from the Verilog library of reversible gates. The library currently supports the design of reversible binary and BCD adders, reversible subtractors and reversible adder-subtractor. The library would be enhanced in future with Verilog codes of design methodologies of various reversible arithmetic units proposed by the authors as well as the other researchers. Thus, in this step, depending on the design requirement and the scheme, the tool selects the Verilog code of the design from the built-in library.

8.2   Code Generation

In this step, the Verilog code of the desired design is generated. The test benches needed to verify the functional correctness of the design are also generated.

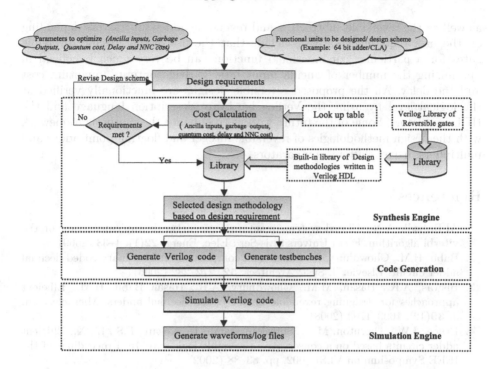

Fig. 20. Proposed Synthesis Framework for design of reversible arithmetic and logic circuits

8.3   Simulation Engine

In this step, the functional verification of our Verilog HDL codes are done using standard HDL simulators such as ModelSim and SynaptiCAD simulators. The waveforms will be generated along with the log files which will have the various costs of the design in terms of number of ancilla inputs, number of garbage outputs, quantum cost and delay.

9   Discussions and Conclusions

In this work, we have presented three different designs methodologies of reversible borrow subtractor and reversible adder-subtractor primarily optimizing the parameters of number of ancilla input bits, the number of garbage outputs, quantum cost and the delay. The proposed reversible subtractors and reversible adder-subtractor designs illustrates the needs of the research towards synthesis of reversible arithmetic circuits as the synthesis of reversible arithmetic circuits is completely different from the synthesis of combinational and sequential circuits. Thus, a synthesis framework to generate Verilog HDL codes of design methodologies of reversible arithmetic units is also proposed. The synthesis framework is currently in preliminary stages and currently supports the automatic generation of Verilog HDL codes of reversible binary and BCD adders as well as the reversible subtractors and reversible adder-subtractors depending on the user requirements. We conclude that the use of the specific reversible gates for a particular combinational function can be very much beneficial in minimizing the number of ancilla input bits, garbage outputs, quantum cost and the delay. All the proposed reversible designs are functionally verified at the logical level by using the Verilog hardware description language and the HDL simulators. Some of the future work is to enhance the synthesis framework with the design methodologies of reversible integer and floating point adder and multipliers, barrel shifters, comparators, etc..

References

1. Al-Rabadi, A.N.: Closed-system quantum logic network implementation of the viterbi algorithm. Facta Universitatis-Ser.: Elec. Energ. **22**(1), 1–33 (2009)
2. Babu, H.M., Chowdhury, A.: Design of a compact reversible binary coded decimal adder circuit. Elsevier J. Syst. Architect. **52**, 272–282 (2006)
3. Biswas, A.K., Hasan, M.M., Chowdhury, A.R., Hasan Babu, H.M.: Efficient approaches for designing reversible binary coded decimal adders. Microelectron. J. **39**(12), 1693–1703 (2008)
4. Bruce, J.W., Thornton, M.A., Shivakumaraiah, L., Kokate, P.S., Li, X.: Efficient adder circuits based on a conservative reversible logic gate. In: Proceedings of the IEEE Symposium on VLSI 2002, pp. 83–88 (2002)
5. Cheng, K.W., Tseng, C.C.: Quantum full adder and subtractor. Electron. Lett. **38**(22), 1343–1344 (2002)
6. Choi, B.S., Van Meter, R.: On the effect of quantum interaction distance on quantum addition circuits. J. Emerg. Technol. Comput. Syst. **7**, 11:1–11:17 (2011). http://doi.acm.org/10.1145/2000502.2000504
7. Chuang, M.L., Wang, C.Y.: Synthesis of reversible sequential elements. J. Emerg. Technol. Comput. Syst. **3**(4), 1–19 (2008)
8. Cuccaro, S.A., Draper, T.G., Kutin, S.A., Moulton, D.P.: A new quantum ripple-carry addition circuit, October 2004. http://arXiv.org/quant-ph/0410184
9. Desoete, B., Vos, A.D.: A reversible carry-look-ahead adder using control gates. Integr. VLSI J. **33**(1), 89–104 (2002)
10. Donald, J., Jha, N.K.: Reversible logic synthesis with fredkin and peres gates. J. Emerg. Technol. Comput. Syst. **4**, 2:1–2:19 (2008)
11. Fredkin, E., Toffoli, T.: Conservative logic. Int. J. Theor Phys. **21**, 219–253 (1982)
12. Golubitsky, O., Maslov, D.: A study of optimal 4-bit reversible toffoli circuits and their synthesis. IEEE Trans. Comput. **61**(9), 1341–1353 (2012)
13. Gupta, P., Agarwal, A., Jha, N.K.: An algorithm for synthesis of reversible logic ciruits. IEEE Trans. Comput. Aided Des. **25**(11), 2317–2330 (2006)
14. Yang, G., Song, X., Hung, W.N., Perkowski, M.: Bi-directional synthesis of 4-bit reversible circuits. Comput. J. **51**(2), 207–215 (2008)
15. Thapliyal, H., Ranganathan, N.: Design of reversible sequential circuits optimizing quantum cost, delay and garbage outputs. ACM J. Emerg. Technol. Comput. Syst. **6**(4), 14:1–14:35 (2010)
16. Haghparast, M., Jassbi, S., Navi, K., Hashemipour, O.: Design of a novel reversible multiplier circuit using HNG gate in nanotechnology. World App. Sci. J. **3**(6), 974–978 (2008)

17. Hung, W.N., Song, X., Yang, G., Yang, J., Perkowski, M.: Optimal synthesis of multiple output boolean functions using a set of quantum gates by symbolic reachability analysis. IEEE Trans. Comput. Aided Des. **25**(9), 1652–1663 (2006)
18. James, R.K., Jacob, K.P., Sasi, S.: Reversible binary coded decimal adders using toffoli gates. In: Ao, S.-L., Rieger, B., Chen, S.-S. (eds.) Advances in Computational Algorithms and Data Analysis. LNEE, vol. 15, pp. 117–131. Springer, Heidelberg (2008)
19. Khan, M.: Design of full-adder with reversible gates. In: Proceedings of the International Conference on Computer and Information Technology, pp. 515–519 (2002)
20. Kotiyal, S., Thapliyal, H., Ranganathan, N.: Reversible logic based multiplication computing unit using binary tree data structure. J. Supercomputing **71**(7), 1–26 (2015)
21. Khan, M.H.A., Perkowski, M.A.: Quantum ternary parallel adder/subtractor with partially-look-ahead carry. J. Syst. Architect. **53**(7), 453–464 (2007)
22. Maslov, D., Dueck, G.W.: Reversible cascades with minimal garbage. IEEE Trans. Comput. Aided Des. **23**(11), 1497–1509 (2004)
23. Maslov, D., Dueck, G., Miller, D., Negrevergne, C.: Quantum circuit simplification and level compaction. IEEE Trans. Comput. Aided Des. Integr. Circuits Syst. **27**(3), 436–444 (2008)
24. Maslov, D., Miller, D.M.: Comparison of the cost metrics for reversible and quantum logic synthesis (2006). http://arxiv.org/abs/quant-ph/0511008
25. Maslov, D., Saeedi, M.: Reversible circuit optimization via leaving the boolean domain. IEEE Trans. Comput. Aided Des. Integr. Circuits Syst. **30**(6), 806–816 (2011)
26. Matsunaga, T., Matsunaga, Y.: Customizable framework for arithmetic synthesis. In: Proceedings of the 12th Workshop on Synthesis And System Integration of Mixed Information technologies (SASIMI 2004), Yokahama, pp. 315–318 (2004)
27. Meter, R., Munro, W., Nemoto, K., Itoh, K.M.: Arithmetic on a distributed-memory quantum multicomputer (2009). http://arxiv.org/abs/quant-ph/0607160
28. Mohammadi, M., Eshghi, M.: On figures of merit in reversible and quantum logic designs. Quantum Inf. Process. **8**(4), 297–318 (2009)
29. Mohammadi, M., Eshghi, M., Haghparast, M., Bahrololoom, A.: Design and optimization of reversible BCD adder/subtractor circuit for quantum and nanotechnology based systems. World Appl. Sci. J. **4**(6), 787–792 (2008)
30. Mohammadi, M., Haghparast, M., Eshghi, M., Navi, K.: Minimization optimization of reversible BCD-full adder/subtractor using genetic algorithm and don't care concept. Int. J. Quantum Inf. **7**(5), 969–989 (2009)
31. Murali, K.V.R.M., Sinha, N., Mahesh, T.S., Levitt, M.H., Ramanathan, K.V., Kumar, A.: Quantum information processing by nuclear magnetic resonance: experimental implementation of half-adder and subtractor operations using an oriented spin-7/2 system. Phys. Rev. A **66**(2), 022313 (2002)
32. Nielsen, M.A., Chuang, I.L.: Quantum Computation and Quantum Information. Cambridge University Press, New York (2000)
33. Oliveira Jr., I., Sarthour, R., Bonagamba, T., Azevedo, E., Freitas, J.C.C.: NMR Quantum Information Processing. Elsevier Science, Amsterdam (2007)
34. Peres, A.: Reversible logic and quantum computers. Phys. Rev. A, Gen. Phys. **32**(6), 3266–3276 (1985)
35. Prasad, A.K., Shende, V., Markov, I., Hayes, J., Patel, K.N.: Data structures and algorithms for simplifying reversible circuits. ACM JETC **2**(4), 277–293 (2006)
36. Rice, J.E.: An introduction to reversible latches. Comput. J. **51**(6), 700–709 (2008)

37. Saeedi, M., Zamani, M.S., Sedighi, M., Sasanian, Z.: Reversible circuit synthesis using a cycle-based approach. J. Emerg. Technol. Comput. Syst. **6**, 131–1326 (2010)
38. Shende, V.V., Prasad, A., Markov, I., Hayes, J.: Synthesis of reversible logic circuits. IEEE Trans. CAD **22**, 710–722 (2003)
39. Sastry, S.K., Shroff, H.S., Mahammad, S.N., Kamakoti, V.: Efficient building blocks for reversible sequential circuit design. In: Proceedings of the 49th IEEE International Midwest Symposium on Circuits and Systems, Puerto Rico, pp. 437–441, August 2006
40. Smolin, J.A., DiVincenzo, D.P.: Five two-bit quantum gates are sufficient to implement the quantum fredkin gate. Phys. Rev. A **53**, 2855–2856 (1996)
41. Takahashi, Y.: Quantum arithmetic circuits: a survey. IEICE Trans. Fundam. **E92–A**(5), 1276–1283 (2010)
42. Takahashi, Y., Kunihiro, N.: A linear-size quantum circuit for addition with no ancillary qubits. Quantum Inf. Comput. **5**(6), 440–448 (2005)
43. Takahashi, Y., Tani, S., Kunihiro, N.: Quantum addition circuits and unbounded fan-out, October 2009. http://arxiv.org/abs/0910.2530
44. Thapliyal, H., Ranganathan, N.: Design of efficient reversible binary subtractors based on a new reversible gate. In: Proceedings of the IEEE Computer Society Annual Symposium on VLSI, Tampa, Florida, pp. 229–234, May 2009
45. Thapliyal, H., Ranganathan, N.: Reversible logic-based concurrently testable latches for molecular QCA. IEEE Trans. Nanotechnol. **9**(1), 62–69 (2010)
46. Thapliyal, H., Ranganathan, N., Ferreira, R.: Design of a comparator tree based on reversible logic. In: Proceedings of the 10th IEEE International Conference on Nanotechnology, Seoul, Korea, pp. 1113–1116, August 2010
47. Thapliyal, H., Ranganathan, N.: Design of efficient reversible logic-based binary and BCD adder circuits. ACM J. Emerg. Technol. Comput. Syst. (JETC) **9**(3), 17 (2013)
48. Thomsen, M., Glück, R.: Optimized reversible binary-coded decimal adders. J. Syst. Archit. **54**(7), 697–706 (2008)
49. Toffoli, T.: Reversible computing. Technical Report, Tech memo MIT/LCS/TM-151, MIT Lab for Computer Science (1980)
50. Trisetyarso, A., Meter, R.V.: Circuit design for a measurement-based quantum carry-lookahead adder (2009). http://arxiv.org/abs/0903.0748
51. Vedral, V., Barenco, A., Ekert, A.: Quantum networks for elementary arithmetic operations. Phys. Rev. A **54**(1), 147–153 (1996)
52. Kotiyal, S., Thapliyal, H., Ranganathan, N.: Efficient reversible NOR gates and their mapping in optical computing domain. Microelectron. J. **6**, 825–834 (2014)

Balancing Load on a Multiprocessor System with Event-Driven Approach

Alexander Degtyarev$^{(\boxtimes)}$ and Ivan Gankevich

Saint Petersburg State University, Saint Petersburg 199034, Russia
deg@csa.ru, i.gankevich@spbu.ru

Abstract. There are many causes of imbalanced load on a multiprocessor system such as heterogeneity of processors, parallel execution of tasks of varying complexity and also difficulties in estimating complexity of a particular task, however, if one can treat computer as an event-driven processing system and treat tasks as events running through this system the problem of load balance can be reduced to a well-posed mathematical problem which further simplifies to solving a single equation. The load balancer measures both complexity of the task being solved and performance of a computer running this particular task so that a load distribution can be adjusted accordingly. Such load balancer is implemented as a computer program and is known to be able to balance the load on heterogeneous processors in a number of scenarios.

Keywords: Load balance · Event-driven architecture · Heterogeneous system · Multiprocessor computer

Introduction

Load balance is maintained by adjusting distribution of computational tasks among available processors with respect to their performance, and inability to distribute them evenly stems not only from technical reasons but also from peculiarity of a problem being solved. On one hand, load imbalance can be caused by heterogeneity of the tasks and inability to estimate how much time it takes to execute one particular task compared to some other task. Such difficulties arise in fluid mechanics applications involving solution of a problem with boundary conditions when the formula used to calculate boundary layer differs from the formula used to calculate inner points and takes longer time to calculate; the same problem arises in concurrent algorithms of intelligent systems that have different asymptotic complexities but solve the same problem concurrently hoping to obtain result by the fastest algorithm. On the other hand, load imbalance can be caused by heterogeneity of the processors and their different performance when solving the same problem and it is relevant when tasks are executed on multiple computers in a network or on a single computer equipped with an accelerator. Therefore, load imbalance can be caused by heterogeneity of tasks and heterogeneity of processors and these peculiarities should be both taken into account to maintain load balance of a computer system.

© Springer-Verlag Berlin Heidelberg 2016
M.L. Gavrilova and C.J. Kenneth Tan (Eds.): Trans. on Comput. Sci. XXVII, LNCS 9570, pp. 35–52, 2016.
DOI: 10.1007/978-3-662-50412-3_3

From mathematical point of view, load balance condition means equality of distribution function F of some task metric (e.g. execution time) to distribution function G of some processor metric (e.g. performance) and the problem of balancing the load is reduced to solving equation

$$F(x) = G(n), \tag{1}$$

where x is task metric (or time taken to execute the task) and n is processor metric (or relative performance of a processor needed to execute this task). Since in general case it is impossible to know in advance neither the time needed to execute the task on a particular processor, nor the performance of a processor executing a particular task, stochastic approach should be employed to estimate those values. Empirical distribution functions can be obtained from execution time samples recorded for each task: task metric is obtained dividing execution time by a number of tasks and processor metric is obtained dividing a number of tasks by their execution time. Also, any other suitable metrics can be used instead of the proposed ones, e.g. the size of data to be processed can be used as a task metric and processor metric can be represented by some fixed number.

It is easy to measure execution time of each task when the whole system acts as an event-driven system and an event is a single task consisting of program code to be executed and data to be processed. In this interpretation, load balancer component is connected to a processor recursively via profiler forming feedback control system. Profiler collects execution time samples and load balancer estimates empirical distribution functions and distributes new tasks among processors solving Eq. 1.

Static load balancing is also possible in this event-driven system and for that purpose a set of different load balancers can be composed into a hierarchy. In such hierarchy, distribution function is estimated incrementally from bottom to top and hierarchy is used to maintain static load balance. Physical processors are composed or decomposed into virtual ones grouping a set of processors and assigning them to a single load balancer or assigning one physical processor to more than one load balancer at once. So static load balancing is orthogonal to dynamic load balancing and they can be used in conjunction.

To summarize, recursive load balancing approach targets problems exhibiting not only dynamic but also static imbalance and the balance can be achieved solving a single equation.

Related Work

The main drawback of existing parallel programming technologies is their inability to perform load balancing across different computing devices. Each device is associated with a different type of a workload, e.g. disk is associated with I/O and processor with pure computations. Although, almost any program involves computations and reading/writing data to disk, today's standards for multi-core programming (like OpenMP [2]) do not allow to do it in parallel — there are no pipelines neither in OpenMP standard nor in any emerging technology known to date. Moreover, there are many other computing devices that can benefit from pipelines — mainly network interface cards and GPUs — to perform computations and data transfer simultaneously. So, modern parallel programming technologies do not allow to co-exist different type of workloads in a single program, but many programs may benefit from it, exploiting additional degrees of parallelism.

In contrast to parallel programming technologies, event-driven approach allows to use every device in the computer in a unified way, and easily form a pipeline between different devices. Event-driven architecture have been used extensively to create desktop applications with graphical user interface since MVC paradigm [11] was developed and nowadays it is also used to compose enterprise application components into a unified system with message queues [9, 13], however, it is rarely implemented in scientific applications. One example of such usage is GotoBLAS2 library [7, 8]. Although, it is not clear from the referenced papers, analysis of its source code[1] shows that this library uses specialized server object to schedule matrix and vector operations' kernels and to compute them in parallel. The total number of CPUs is defined at compile time and they are assumed to be homogeneous. There is a notion of a queue implemented as a linked list of objects where each object specifies a routine to be executed and data to be processed and also a number of CPUs to execute it on. Server processes these objects in parallel and each kernel can be executed in synchronous (blocking) and asynchronous (non-blocking) mode. So, compared to event-driven system GotoBLAS2 server uses static task scheduling, its tasks are not differentiated into production and reduction tasks, both the tasks and the underlying system are assumed to be homogeneous. GotoBLAS2 library exhibits competitive performance compared to other BLAS implementations [7, 8] and it is a good example of viability of event-driven approach in scientific applications. Considering this, event-driven system can be seen as a generalization of this approach to a broader set of applications.

There are a number of research works in which the authors develop systems borrowing some features from event-driven approach. For example, in [10, 12, 14] a concurrent object-oriented system similar to our system is described. In contrast to our system, it uses messages rather than events to transfer data, and it does not allow load-balancing across different computing devices. Another example is [1], in which the author describes a system which uses *supervisor trees* to organize concurrent processes and manage resources. These structures are similar to hierarchies, but we have different hierarchies for servers and tasks, so that system resources and the flow of computations can be managed separately. So, there are some works which borrow different parts of a typical event-driven system, however, they either do not put emphasis on load-balancing, or do not describe their system as an event-driven, thus not exploiting full benefits of it.

Load-balancing is one of the main tasks of an operating system, and in our view in high-performance computing it should be delegated to some intermediate software layer which lies between operating system and application. So, in contrast to

---

[1] Source code is available in https://www.tacc.utexas.edu/tacc-projects/gotoblas2/.

described approaches it would be useful to implement full event-driven system and hide message passing and synchronization logic in it.

1 Implementation of Event-Driven System

The whole system was implemented as a collection of C++ classes, and problem-solving classes were separated from utility classes with an event-driven approach. In this approach, problem solution is represented by a set of executable objects or "employees", each implementing a solution of one particular part of a problem. Each executable object can implement two methods. With the first method employee either solves part of the problem or produces child executable objects (or "hires" additional employees) to delegate problem solution to them. Since upon completion of this method no object is destroyed, it is called "production" task or "upstream" task as it often delegates problem solution to a hierarchy layer located farther from the root than the current layer is. The second method collects execution results from subordinate executable objects and takes such object as an argument. Upon completion of this method the child object is destroyed or "fired" so that the total number of executable objects is reduced. Hence this task is called "reduction" task or "downstream" task since the results are sent to a hierarchy layer located closer to root. Executable objects can send results not only to their parents but also to any number of other executable objects, however, when communication with a parent occurs the child object is destroyed and when the root object tries to send results to its nonexistent parent, the program ends. Execution of a particular object is performed via submitting it to a queue corresponding to a particular processor. Child and parent objects are determined implicitly during submission so that no manual specification is needed. Finally, these objects are never copied and are accessed only via their addresses. In other words, the only thing that is required when constructing an executable object is to implement a specific method to solve a task and object's life time is implicitly controlled by the system and a programmer does not have to manage it manually.

Execution of objects is carried out concurrently and construction of an executable object is separated from its execution with a thread-safe queue. Every message in a queue is an executable object and carries the data and the code needed to process it and since executable objects are completely independent of each other they can be executed in any order. There are real server objects corresponding to each queue in a system which continuously retrieve objects from a queue and execute their production or reduction tasks in a thread associated with the server object. Production tasks can be submitted to any queue, but a queue into which reduction tasks can be submitted is determined by a corresponding parent object so that no race condition can occur. Since each processor works with its own queue only and in its own thread, processing of queues is carried out concurrently. Also, each queue in a system represents a pipeline through which the data flows, however, execution order is completely determined by the objects themselves. So, executable objects and their methods model control flow while queues model data flow and the flows are separated from each other.

Heterogeneity of executable objects can cause load imbalance among different queues and this problem can be solved introducing imaginary (i.e. proxy) servers and profilers to aid in distribution of executable objects. Imaginary server is a server tied to a set of other servers and its only purpose is to choose the right child server to execute an object at.

In the simplest case, a proper distribution can be achieved with round-robin algorithm, i.e. when each arriving object is executed on the next server, however, in general case, some additional information about completed runs is needed to choose the right server and this information can be collected with pluggable profiler objects. When a new object arrives to an imaginary server, actual profiling information is collected from child servers and specified distribution strategy is used to delegate execution of an object to an appropriate server, and some static distribution strategy is also possible. So, imaginary servers together with distribution strategies and profilers can be used to distribute executable objects among real servers taking into account some profiling information of completed object executions.

The class diagram of the whole event-driven system is depicted in the Fig. 1 and the system works as follows.

1. When a program execution starts, the hierarchy of imaginary and real servers is composed. All real servers are launched in a separate threads and processing of executable objects starts.
2. The first object is created and submitted to the imaginary server at the top of the hierarchy. The server employs specified distribution strategy to choose an appropriate server from the next layer of the hierarchy to send the object to. The profiler gathers measurements of completed runs from subordinate servers and decides where to send an object.
3. The previous step repeats until the bottom level of the hierarchy is reached and real server which was found with the distribution strategy starts execution of an object.
4. Object is executed and measurements are made by a profiler. If during execution more executable objects are created and submitted to the top imaginary server, the whole algorithm is repeated for each new object; if the root object submits reduction task then all servers in the hierarchy are shut down, and program execution ends.

To sum up, the whole system is composed of the two hierarchies: one hierarchy represents tasks and data and their dependencies employing executable objects, the other hierarchy represents processing system employing imaginary and real server objects. Mapping of the first hierarchy to the second is implicit and is implemented using message queues. Such composition allows easy configuration of dynamic and static load distribution strategies and allows programming with simple executable objects.

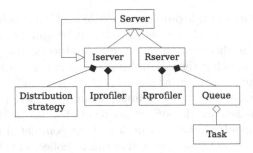

Fig. 1. Class diagram for an event-driven system. "Iserver" denotes imaginary server and "Rserver" denotes real server.

2 Implementation of Distribution Strategy

Recursive load balancing was implemented as a load distribution strategy, however, Eq. 1 was not solved directly. The first problem occurring when solving this equation directly was that task metric x cannot be computed before actually running the task so it was estimated to be an average metric of a number of previous runs. The second problem was that when the task metric is known, the result of direct solution of Eq. 1 is not an identifier of a processor to execute the task on but it is number n – relative performance of a processor needed to execute the task and the number n is not particularly useful when determining where to execute the task. Therefore, Eq. 1 was not solved directly but its main idea was realized in an algorithm similar to round-robin.

The resulting algorithm works as follows.

1. First, algorithm collects samples recorded by profilers of child servers as well as estimates task metric and processor metric using values from previous runs. At this stage, not only averaging but also any other suitable predicting technique can be used.
2. Then, probability of having a task with metric equal to computed task metric is determined by counting samples equal to computed task metric and dividing it by the total number of samples.
3. The cursor pointing at the processor to execute the next task on is incremented by a step equal to a product of computed probability and computed processor metric.
4. Then, by recursively subtracting metric of each processor from the cursor, the needed processor is found and the task is executed on it.

The resulting mathematical formula for each step can be written simply as

$$cursor = cursor + F(\bar{x})\bar{n},$$

where $\bar{x}$ is a task metric and $\bar{n}$ is a processor metric. In case of fully homogeneous system and all tasks having equal metric this algorithm is equivalent to round-robin: all processors have metric equal to 1 and probability is always 1 so that the cursor is always incremented by 1.

Although, the algorithm is simple, in practice it requires certain modifications and a robust profiler to work properly. Since algorithm balances reciprocal values of task metric t (execution time) and processor metric $1/t$ (processor throughput), even a slight oscillation of a task metric can affect the resulting distributions greatly. The solution to this problem is to smooth samples with a logarithm function and it can be done in a straightforward way, because the algorithm does not make assumptions about metrics' dimensions and treat them as numbers. The second problem is that the algorithm should be implemented with integer arithmetic only to minimize overhead of load balancing. This problem can be solved by omitting mantissa after logarithm is applied to a sample and in that case processor metric is equal to task metric but has an opposite sign. The last major problem is that the distribution of task metric may change abruptly during program execution, which renders samples collected by a profiler for previous runs useless. This problem is solved by detecting a sharp change in task execution time (more than three standard deviations) and when outliers are detected the profiler is reset to its initial state in which distribution is assumed to be uniform. As a result of applying logarithm to each sample the algorithm becomes unsuitable for relatively small tasks and for tasks taking too much time, and although such tasks are executed, the samples are not collected for them as they often represent just control flow tasks. To summarize, the modified algorithm is implemented using integer arithmetic only, is suitable for relatively complex tasks and adapts to a rapid change of a task metric distribution.

One problem of the algorithm that stands aside is that it becomes inefficient in the event of high number of tasks with high metric values. It happens because when task is assigned to a particular processor it is not executed directly but rather gets placed in a queue. If this queue is not empty the task can reside in it for such a long time that its assignment to a particular processor will not match actual distribution function. The solution is simple: these stale tasks can be easily detected by recording their arrival time and comparing it with the current time and when such tasks are encountered by a queue processor, they can be redistributed to match the current distribution function. However, an existence of stale tasks is also an evidence that the computer is not capable of solving a problem fast enough to cope with continuously generated tasks and it is an opportunity to communicate with some other computers to solve the problem together. From a technical point of view, delegation of tasks to other computers is possible because tasks are independent of each other and read/write (serialization) methods are easily implemented for each of them, however, the problem was not addressed herein, and only load redistribution within a single computer was implemented.

Described algorithm is suitable for distributing production tasks, but a different algorithm is needed to distribute reduction tasks. Indeed, when executable objects come in pairs consisting of the child and its parent, all children of the parent must be executed on the same server so that no race condition takes place, so it is not possible to distribute the task on an arbitrary server but a particular server must be chosen for all of the child tasks. One possible way of choosing a server is by applying a simple hashing function to parent's memory address. Some sophistication of this algorithm is possible, e.g. predicting memory allocation and deallocation pattern to distribute reduction tasks uniformly among servers, however, considering that most of the reduction tasks in tested program were simple (the reason for this is discussed in Sect. 3.1) the approach seemed to be non viable and was not implemented. So, simple hashing algorithm was used to distribute reduction tasks among servers.

To summarize, recursive load distribution algorithm by default works as round-robin algorithm and when a reasonable change of task execution time is detected it automatically distributes the load in accordance with task metric distribution. Also, if there is a change in processor performance it is taken into account by relating its performance to other processors of computing system. Finally, if a task stays too much time in a queue it is distributed once again to match current distribution function.

3   Evaluation

Event-driven approach was tested on the example of hydrodynamics simulation program which solves real-world problem [3–6]. The problem consists of generating real ocean wavy surface and computing pressure under this surface to measure impact of the external excitations on marine object. The program is well-balanced in terms of processor load and for the purpose of evaluation it was implemented with introduced event-driven approach and the resulting implementation was compared to existing non event-driven approach in terms of performance and programming effort.

Event-driven architecture makes it easy to write logs which in turn can be used to make visualization of control flow in a program. Each server maintains its own log file and when some event occurs it is logged in this file accompanied by a time stamp and a server identifier. Having such files available, it is straightforward to reconstruct a sequence of events occurring during program execution and to establish connections between these events (to dynamically draw graph of tasks as they are executed). Many such graphs are used in this section to demonstrate results of experiments.

Generation of a wavy surface is implemented as a transformation of white noise, autoregressive model is used to generate ocean waves and pressures are computed using analytical formula. The program consists of preprocessing phase, main computer-intensive phase and post-processing phase. The program begins with solving Yule-Walker equations to determine autoregressive coefficients and variance of white noise. Then white noise is generated and is transformed to a wavy surface. Finally, the surface is trimmed and written to output stream. Generation of a wavy surface is the most computer-intensive phase and consumes over 80 % of program execution time (Fig. 7) for moderate wavy surface sizes and this time does not scale with a surface size. So, the program spends most the time in the main phase generating wavy surface (this phase is marked with $[G_0, G_1]$

interval in the graphs). The hardware used in the experiments is listed in Table 2. The program was tested in a number of experiments and finally compared to other parallel programming techniques.

3.1   Evaluation of Event-Driven System

The first experiment consisted of measuring stale cycles and discovering causes of their occurrence. Program source code was instrumented with profiling directives and every occurrence of stale cycles was written to the log file. Also the total stale time was measured. Obtained results showed that stale cycles prevail in preprocessing and at the end of main phase but are not present in other parts of the program (Fig. 2). The reason for this deals with insufficient amount of tasks available to solve during these phases which in turn is caused by global synchronizations occurring multiple times in preprocessing phase and naturally at the end of a program. Stale cycles in the main phase are caused by computation performing faster than writing results to disk: in the program only one thread writes data and no parallel file system is used. Further experiments showed that stale cycles consume at most 20 % of the total execution time for 4 core system (Table 1) and although during this time threads are waiting on a mutex so that this time can be consumed by other operating system processes, there is also an opportunity to speed up the program. Considering file output performance stale cycles can only be reduced with faster storage devices combined with slower processors or with parallel file systems combined with fast network devices and interconnects. In contrast, the main cause of stale cycles in preprocessing phase deals with global synchronization and to minimize its effect it should be replaced by incremental synchronization if possible.

The next experiment consisted of measuring different types of overheads including profiling, load balancing, queuing and other overheads so that real performance of event-driven system can be estimated. In this experiment, the same technique was used to obtain measurements: every function causing overhead was instrumented and also the total time spent executing tasks and total program execution time was measured. As a result, the total overhead was estimated to

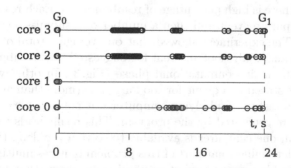

Fig. 2. Occurrences of stale cycles in preprocessing and at the end of the main computational phase of a program. Range $[G_0, G_1]$ denotes computationally intensive phase.

be less than 0.1 % for different number of cores (Table 1). Also the results showed that reduction time is smaller than the total time spent solving production tasks in all cases (Table 1). It is typical of generator programs to spend more time solving data generating production tasks than solving data processing reduction tasks; in a data-centric program specializing in data processing this relation can be different. Finally, it is evident from the results that the more cores are present in the system the more stale time is introduced into the program. This behavior was explained in the previous experiment and is caused by imbalance between processor performance and performance of a storage device for this particular computational problem. To summarize, the experiment showed that event-driven system and recursive load distribution strategy do not incur much overhead even on systems with large number of cores and the program is rather code-centric than data-centric spending most of its execution time solving production tasks.

Table 1. Distribution of wall clock time and its main consumers in event-driven system. Time is shown as a percentage of the total program execution time. Experiments for 4 cores were conducted on the system I and experiments for 24 and 48 cores were conducted on the system II from Table 2.

Classifier	Time consumer	Time spent, %		
		4 cores	24 cores	48 cores
Problem solution	Production tasks	71	33	19
	Reduction tasks	13	4	2
Stale time	Stale cycles	16	63	79
Overhead	Load distribution overhead	0.01	0.0014	0.0017
	Queuing overhead	0.002	0.0007	0.0005
	Profiling overhead	0.0004	0.0004	0.0003
	Other overheads	0.06	0.03	0.02

In the third experiment, the total number of production tasks solved by the system was measured along with the total number of task resubmissions and it was found that there is high percentage of resubmissions. Each resubmission was recorded as a separate event and then a number of resubmissions for each task was calculated. The experiment showed that on average a total of 35 % of tasks are resubmitted and analysis of an event log suggested that resubmissions occur mostly during the main computational phase (Fig. 3). In other words 35 % of production tasks stayed in a queue for too long time (more than an average time needed to solve a task) so underlying computer was not capable of solving tasks as fast as they are generated by the program. This result leads to a conclusion that if more than one computer is available to solve a problem, then there is a natural way to determine what part of this problem requires multiple computers to be solved. So, high percentage of resubmissions shows that machine solves production tasks slower than they are generated by the program so multiple machines can be used to speed up problem solution.

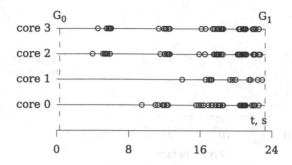

Fig. 3. Event plot of resubmission of production tasks staying in a queue for too long time. Range $[G_0, G_1]$ denotes computationally intensive phase.

In the final experiment overall performance of event-driven approach was tested and it was found to be superior when solving problems producing large volumes of data. In the previous research it was found that OpenMP is the best performing technology for the wavy ocean surface generation [3], so the experiment consisted of comparing its performance to the performance of event-driven approach on a set of input data. A range of sizes of a wavy surface was the only parameter that was varied among subsequent program runs. As a result of the experiment, event-driven approach was found to have higher performance than OpenMP technology and the more the size of the problem is the bigger performance gap becomes (Fig. 4). Also event plot in Fig. 5 of the run with the largest problem size shows that high performance is achieved with overlapping of parallel computation of a wavy surface (interval $[G_0, G_1]$) and output of resulting wavy surface parts to the storage device (interval $[W_0, W_1]$). It can be seen that there is no such overlap in OpenMP implementation and output begins at point W_0 right after the generation of wavy surface ends at point G_1. In contrast, there is a significant overlap in event-driven implementation and in that case wavy surface generation and data output end almost simultaneously at points G_1 and W_1 respectively. So, approach with pipelined execution of parallelized computational steps achieves better performance than sequential execution of the same steps.

Although OpenMP technology allows constructing pipelines, it is not easy to combine a pipeline with parallel execution of tasks. In fact such combination is possible if a thread-safe queue is implemented to communicate threads generating ocean surface to a thread writing data to disk. Then using *omp section* work of each thread can be implemented. However, implementation of parallel execution within *omp section* requires support for nesting *omp parallel* directives. So, combining pipeline with parallel execution is complicated in OpenMP implementation requiring the use a thread-safe queue which is not present in OpenMP standard.

To summarize, event-driven programming approach was applied to a real-world high-performance application and it was shown that it incurs low overhead, but results in appearance of stale periods when no problem solving is performed

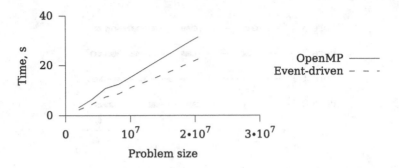

Fig. 4. Performance comparison of OpenMP and event-driven implementations.

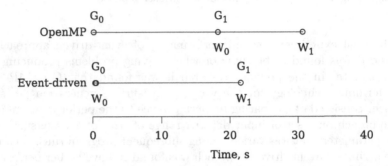

Fig. 5. Event plot showing overlap of parallel computation $[G_0, G_1]$ and data output $[W_0, W_1]$ in event-driven implementation. There is no overlap in OpenMP implementation.

by some threads. The duration of these periods in the main phase can be reduced with faster storage equipment and the duration of stale periods in preprocessing phase can be reduced employing incremental synchronization techniques. Also, event-driven approach offers a natural way of determining whether program execution should scale to multiple machines or not, however, viability of such mode of execution was not tested in the present research. Finally, it was shown that event-driven approach is more efficient than standard OpenMP technology especially for large problem sizes and it was also shown that a pipeline combined with parallel execution works faster than sequential execution of parallelized steps.

3.2   Evaluation of Load Distribution Strategy

Performance of recursive load distribution algorithm was compared to performance of round-robin algorithm and was tested in a number of scenarios with combinations of homogeneous and heterogeneous tasks and homogeneous and heterogeneous processors. In each experiment the total execution time and distributions of task metric and processor metric were measured and compared to uniform distribution case. All tests were performed on the same system (Table 2) and each scenario was run multiple times to ensure accurate results. Also, preliminary validation tests were performed to make sure that the algorithm works as intended. So, the purpose of evaluation was to demonstrate how algorithm works in practice and to measure its efficiency on a real problem.

Table 2. Testbed setup.

Component	System	
Programming language	C++11	
Threading library	C++11 STL threads	
Atomics library	C++11 STL atomic	
Time measurement routines	`clock_gettime(CLOCK_MONOTONIC, ...)`	
	`/usr/bin/time -f %e`	
Compiler	GCC 4.8.2	
Compiler flags	`-std=c++11 -O2 -march=native`	
	I	II
Operating system	Debian 3.2.51-1 x86_64	CentOS 6.5 x86_64
File system	ext4	ext4
Processor	Intel Core 2 Quad Q9650	2×Intel Xeon E5-2695 v2
Cores frequency (GHz)	3.00	2.40
Number of cores	4	24 (48 virtual cores)
RAM capacity (GB)	8	256
RAID device		Dell PERC H710 Mini
RAID configuration		RAID10
Storage device	Seagate ST3250318AS	4×Seagate ST300MM0006
Storage device speed (rpm)	7200	2×10000

It has already been shown that the algorithm consumes only a small fraction of total execution time of a program (Table 1), so the purpose of the validation test was to show algorithm's ability to switch between different task metric distributions. The switching is performed when a significant change (more than three standard deviations) of a task execution time occurs. The test have shown that the switching events are present in preprocessing phase and do not occur in the main phase (Fig. 7). The cause of the switching is a highly variable task execution time inherent to preprocessing phase. So, profilers' resets occur only when a change of task execution time distribution is encountered and no switching is present when this distribution does not change.

The purpose of the first experiment was to show that the algorithm is capable of balancing homogeneous tasks on homogeneous computer and in that case it works like well-known round-robin algorithm. During the experiment, events of task submissions were recorded as well as additional profiler data and an event plot was created. In Fig. 7a relative performance of each processor core is plotted and all the samples lie on a single line in the computational phase. Since this phase consists of executing tasks of equal metric, the straight line represents the uniform distribution of tasks among processor cores constituting round-robin algorithm. So, in the simplest case of homogeneous tasks and processors recursive load balancing algorithm works as round-robin algorithm.

The purpose of the second experiment was to show that recursive load-balancing algorithm is capable of balancing homogeneous tasks on heterogeneous processors and in that case it can distribute the load taking into account performance of a particular processor. Although natural application of such load balancing is hybrid computer systems equipped with graphical or other accelerators, the experiment was conducted by emulating such systems with a hierarchy of servers. It was found that load balancing algorithm can recognize performance of different components and adapt distribution of tasks accordingly (Fig. 7b): I_1's first and second child servers have relative performance equal to 0.75 and 0.25 respectively whereas all children of I_2 server have relative performance equal to $\frac{1}{3}$. Also, this system setup shows performance similar to performance of the homogeneous computer configuration (Fig. 6). So, recursive load balancing algorithm works on heterogeneous computer configurations and the performance is similar to homogeneous system case.

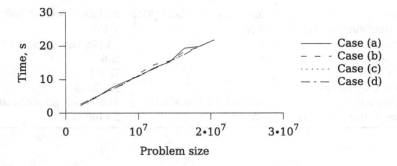

Fig. 6. Performance comparison of different server configurations. Configurations are listed in Fig. 7.

The purpose of the third experiment was to show that the algorithm is capable of balancing heterogeneous tasks on a homogeneous system and the experiment showed that performance gain is small. For the experiment the source code generating a wavy surface was modified so that parts of two different sizes are generated simultaneously. In order to balance such workload on a homogeneous system the step should be equal to $\frac{1}{2i}n, i = 1, 2, ...$, where n is the processor metric (instead of being equal to 1 when parts have the same size) so that each processor takes two respective parts of the surface. In the Fig. 7c showing results of the experiment the step reaches its optimal value of $\frac{1}{2}n$ (0.125 mark), however, it takes almost 8 seconds (or 40 % of the total time) to reach this value. The first two cases do not exhibit such behavior and the step does not change during execution. Also, in the course of the experiment it was found that the

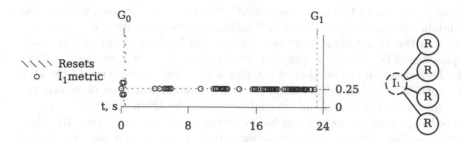

(a) Homogeneous tasks and homogeneous computer case.

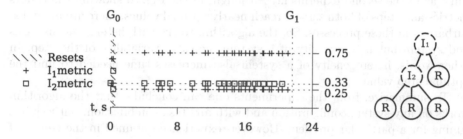

(b) Homogeneous tasks and heterogeneous computer case.

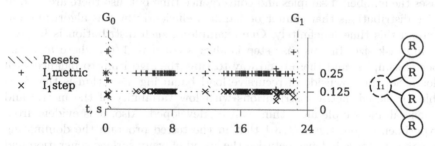

(c) Heterogeneous tasks and homogeneous computer case.

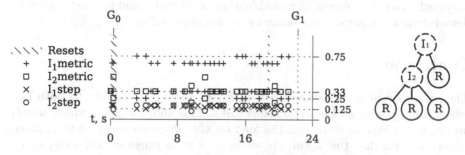

(d) Heterogeneous tasks and heterogeneous computer case.

Fig. 7. Event plot of task submissions and relative performance of child servers recorded at the time of submissions. I denotes "Iserver" and R denotes "Rserver". Profiled servers are marked with dashed line.

step oscillates and to fix this it was smoothed with five point median filter and the number of samples was doubled. Finally, in subsequent experiments it was found that the more unique parts sizes are present in the main phase, the more samples should be collected to preserve the accuracy of the step evaluation, however, the increase in the number of samples led to slow convergence of the step to its optimal value. In other words, the more heterogeneous the tasks are, the more time is needed to find the optimal step value for them.

The purpose of the fourth and final experiment was to show that the algorithm is capable of balancing heterogeneous tasks on heterogeneous system and results were similar to the previous experiment. System configuration was the same as in the second experiment. Although, in the Fig. 7d showing the results metrics and steps of both servers reach nearly optimal values, there are more disturbances in these processes. So, the algorithm works with heterogeneous tasks and system but heterogeneity of a system increases variability of the step. In other words, heterogeneity of a system also increases time needed to find the optimal step value.

To summarize, from the experiments one can conclude, that the algorithm works on any system configuration and with any task combination, but requires tuning for a particular problem. However, experience obtained in the course of the experiments suggests that not only heterogeneity of tasks and computers increases the number of samples and convergence time but also there are certain task size distributions that cannot be handled efficiently by this algorithm and can extend this time indefinitely. One example of such distribution is linearly increasing task size. In this case step is always equal to $1/m$, where m is the number of samples, and there is no way to tune the algorithm to balance such workload. So, the downside of recursive load balancing algorithm is that it is suitable for closed metric distributions with low variability of the metric and more general and simple algorithms can be developed. Also, it is evident from the experiments from the Sect. 3.1 that in the tested program the dominating performance factor is balance between the speed of wavy surface generation and the speed of writing it to storage device. In that case, load balancing algorithm plays only a second role and any combination of computer and task heterogeneity demonstrates comparable performance as was depicted in Fig. 6.

Conclusions

The main advantage of event-driven approach is its applicability to both heterogeneous systems and heterogeneous tasks. This allows a programmer to rely on the technology to distribute the load on the processor cores evenly. Experiments showed that this approach works in a wide range of test cases and a real-world application. Moreover, in this application it performs better than popular OpenMP technology.

Apart from being more efficient than OpenMP the biggest advantage of event-driven approach is the ease of parallel programming. First of all, what is needed from a programmer is to develop a class to describe each independent task, create objects of that class and submit them to a queue. Programming in such a way does not involve thread and lock management and the system is flexible enough to have even the tiniest tasks executed in parallel. Second, relieving programmer from thread management makes it easy to debug this system. Each thread maintains its own log and any of both system and user events can be written to it and the sequence of events can be restored after the execution ends. Finally, with event-driven approach it is easy to write load distribution algorithm for your specific problem (or use an existing one). The only thing which is not done automatically is decomposition and composition of tasks, however, this problem requires higher layer of abstraction to solve.

The future work is to extend event-driven approach for distributed and hybrid (GPGPU) systems and to see if it is possible to cover those cases. The other possible direction of research is to create declarative language which acts as higher layer of abstraction and performs decomposition into tasks automatically.

Acknowledgments. Research was carried out using computational resources provided by Resource Center "Computer Center of SPbU"(Official web site: http://cc.spbu.ru.) and supported by Russian Foundation for Basic Research (project N 13-07-00747) and Saint Petersburg State University (projects N 9.38.674.2013, 0.37.155.2014).

References

1. Armstrong, J.: Making reliable distributed systems in the presence of sodware errors. Ph.D. thesis, The Royal Institute of Technology Stockholm, Sweden (2003)
2. Dagum, L., Enon, R.: Openmp: an industry standard api for shared-memory programming. Computat. Sci. Eng. IEEE **5**(1), 46–55 (1998)
3. Degtyarev, A., Gankevich, I.: Wave surface generation using OpenCL, OpenMP and MPI. In: Proceedings of 8th International Conference "Computer Science & Information Technologies", pp. 248–251 (2011)
4. Degtyarev, A., Gankevich, I.: Evaluation of hydrodynamic pressures for autoregression model of irregular waves. In: Proceedings of 11th International Conference "Stability of Ships and Ocean Vehicles", Athens, pp. 841–852 (2012)
5. Degtyarev, A.B., Reed, A.M.: Modelling of incident waves near the ship's hull (application of autoregressive approach in problems of simulation of rough seas). In: Proceedings of the 12th International Ship Stability Workshop (2011)
6. Degtyarev, A.B., Reed, A.M.: Synoptic and short-term modeling of ocean waves. In: Proceedings of 29th Symposium on Naval Hydrodynamics (2012)
7. Goto, K., Van De Geijn, R.: Anatomy of high-performance matrix multiplication. ACM Trans. Math. Softw. (TOMS) **34**(3), 12 (2008)
8. Goto, K., Van De Geijn, R.: High-performance implementation of the level-3 blas. ACM Trans. Math. Softw. (TOMS) **35**(1), 4 (2008)
9. Hapner, M., Burridge, R., Sharma, R., Fialli, J., Stout, K.: Java Message Service. Sun Microsystems Inc., Santa Clara (2002)
10. Kale, L.V., Krishnan, S.: CHARM++: A Portable Concurrent Object Oriented System Based On C++, vol. 28. ACM, Seattle (1993)
11. Krasner, G.E., Pope, S.T., et al.: A description of the model-view-controller user interface paradigm in the smalltalk-80 system. J. Object Oriented Program. **1**(3), 26–49 (1988)

12. Pilla, L.L., Ribeiro, C.P., Cordeiro, D., Méhaut, J.-F.: Charm++ on numa platforms: the impact of smp optimizations and a numa-aware load balancer. In: 4th Workshop of the INRIA-Illinois Joint Laboratory on Petascale Computing, Urbana, IL, USA (2010)
13. Vinoski, S.: Advanced message queuing protocol. Internet Comput. IEEE **10**(6), 87–89 (2006)
14. Zheng, G., Meneses, E., Bhatele, A., Kale, L.V.: Hierarchical load balancing for charm++ applications on large supercomputers. In: 39th International Conference on Parallel Processing Workshops (ICPPW), pp. 436–444. IEEE (2010)

On Stability of Difference Schemes for a Class of Nonlinear Switched Systems

Alexander Aleksandrov[1(✉)], Alexey Platonov[1], and Yangzhou Chen[2]

[1] Saint Petersburg State University, Saint Petersburg, Russia
{a.u.aleksandrov,a.platonov}@spbu.ru
[2] Beijing University of Technology, Beijing, China
yzchen@bjut.edu.cn

Abstract. The problem of preservation of stability under discretization is studied. A class of nonlinear switched difference systems is considered. Systems of the class appear as computational schemes for continuous-time switched systems with homogeneous right-hand sides. By using the Lyapunov direct method, some sufficient conditions of the asymptotic stability of solutions for difference systems are obtained. These conditions depend on the information available about the switching law. Three cases are considered. In the first case, we can guarantee the asymptotic stability for any switching law, while in the second and in the third ones, classes of switched signals are determined for which the preservation of the asymptotic stability takes place.

Keywords: Switched difference systems · Computational schemes · Stability · Lyapunov functions · Dwell-time

1 Introduction

Constructing conservative computational schemes is fundamental and challenging research problem. Such schemes, nowadays, are playing an important role in the design of reliable numerical methods in several areas in Science and Engineering, such as, chemical kinetics, fluid mechanics, photo-conductivity, semiconductor devices, weather-forecasting, mathematical biology, etc., see [1–5].

Many real life phenomena and processes are modeled by differential equations, for which analytical solutions are not always easy to find [1,2]. Therefore, we often have to visualize the solutions by using discretization methods and computer programming. After applying a discretization (or numerical) method to a system of ordinary differential equations, it produces a system of difference equations, or a discrete-time system [1–3].

Preserving qualitative characteristics (integrals, integral invariants, stability, etc.), when passing from differential equations to finite-difference ones, is an important problem to study [2,4,6–9]. It is well known that in some cases it is necessary to modify numerical schemes in order to preserve the properties. Such modifications lead to conservative numerical schemes that are based

© Springer-Verlag Berlin Heidelberg 2016
M.L. Gavrilova and C.J. Kenneth Tan (Eds.): Trans. on Comput. Sci. XXVII, LNCS 9570, pp. 53–67, 2016.
DOI: 10.1007/978-3-662-50412-3_4

on application of control in the computational process [4,7,8,10]. However, the use of such schemes seriously complicates the resulting difference systems (see [2,8,11]). From a practical point of view, it is important to describe classes of systems for which discretization allows to preserve qualitative characteristics without modification of the computational scheme.

The problem of preserving of qualitative characteristics is essentially complicated in the case of discretization of switched systems. A switched system is a particular kind of hybrid system that consists of a family of subsystems and a switching signal selecting at each time instant which subsystem is active. Such systems have extensive applications in mechanics, automotive industry, aircraft and air traffic control, chemical and electrical engineering, etc. [12–14]. Stability analysis of switched systems is fundamental and challenging research problem, and various approaches have been proposed (see, for example, [12–19]).

In this paper, a special class of switched nonlinear difference systems is considered. These systems appear in numerical simulation of continuous-time switched homogeneous systems by means of the Euler method [1,3]. The problem of preservation of stability under the discretization is studied. By the usage of the Lyapunov direct method, some sufficient conditions of asymptotic stability of the zero solution for difference systems are obtained. These conditions depend on the information available about the switching law. Three cases are considered. In the first case, we can guarantee the asymptotic stability for any switching law, while in the second and in the third ones, classes of switched signals are determined for which the preservation of the asymptotic stability takes place.

2 Statement of the Problem

Let the nonlinear switched system

$$\dot{\mathbf{z}} = \mathbf{F}^{(\omega)}(\mathbf{z}) \tag{1}$$

be given. Here $\mathbf{z} = (z_1, \ldots, z_n)^T$ is the state vector; $\omega = \omega(t)$ is the piecewise constant function defining the switching law, $\omega(t) : [0, +\infty) \to Q = \{1, \ldots, N\}$. Thus, at each time instant, the operation of (1) is described by one of the subsystems

$$\dot{\mathbf{z}} = \mathbf{F}^{(s)}(\mathbf{z}), \qquad s = 1, \ldots, N. \tag{2}$$

We assume that vector functions $\mathbf{F}^{(1)}(\mathbf{z})$, ..., $\mathbf{F}^{(N)}(\mathbf{z})$ are continuous for $\mathbf{z} \in R^n$ and homogeneous of the order $\mu > 1$, where μ is a rational with odd numerator and denominator. It is evident that system (1) admits the zero solution $\mathbf{z} = \mathbf{0}$. Conditions of asymptotic stability of the solution have been obtained in [18].

In this paper, we deal with preservation of asymptotic stability under the digitization of system (1) by means of the Euler method. Consider the corresponding difference system

$$\mathbf{x}(k+1) = \mathbf{x}(k) + h\mathbf{F}^{(\sigma)}(\mathbf{x}(k)).$$

Here $\mathbf{x}(k) = (x_1(k), \ldots, x_n(k))^T$; $h > 0$ is a digitization step; $\sigma = \sigma(k)$, $k = 0, 1, \ldots$, with $\sigma(k) \in \{1, \ldots, N\}$, defines the switching rule. It is assumed that functions $\mathbf{F}^{(1)}(\mathbf{z})$, ..., $\mathbf{F}^{(N)}(\mathbf{z})$ are the same as in (1).

The results of this paper do not depend of the value of the digitization step h. Therefore, without loss of generality, in what follows we set $h = 1$. Thus, consider the switched difference system

$$\mathbf{x}(k+1) = \mathbf{x}(k) + \mathbf{F}^{(\sigma)}(\mathbf{x}(k)), \tag{3}$$

and the family of subsystems

$$\mathbf{x}(k+1) = \mathbf{x}(k) + \mathbf{F}^{(s)}(\mathbf{x}(k)), \qquad s = 1, \ldots, N, \tag{4}$$

which is a discrete counterpart of the family (2).

Let us determine the conditions providing asymptotic stability of the zero solution of system (3). We shall investigate three different cases depending on assumptions imposed upon the switching law $\sigma(k)$.

3   Construction of a Common Lyapunov Function

First, consider the case when the switched signal is unknown. We will look for conditions under which the zero solution of switched system (3) is asymptotically stable for an arbitrary switching law. The general approach to the problem in such a statement is the construction of a common Lyapunov function for the corresponding family of subsystems [12]. It was effectively used in many papers (see, for instance, [12–14, 17–19]). However, up to now the problem of computation of common Lyapunov functions has not got a constructive solution.

In [17] some theorems on the existence of common Lyapunov functions for families of dynamical systems are proved, and, on the base of these theorems, the constructive procedure was developed for finding such functions for families of linear stationary subsystems. In [18] this approach has been used for the construction of a common Lyapunov function for family of homogeneous differential systems of the form (2). The goal of the present section is extension of the above approach to the family of difference systems (4).

It is worth mentioning that methods of Lyapunov functions construction are better developed for continuous-time systems, than for discrete-time ones. However, in some cases, a Lyapunov function found for a differential system can be applied to the corresponding difference system as well [9, 20, 21]. According to this approach, we will construct Lyapunov functions for subsystems from the family (2) and will use these functions for stability analysis of system (3).

Let the following assumptions be fulfilled:

Assumption 1. The zero solutions of subsystems (2) are asymptotically stable.

Assumption 2. For subsystems (2), Lyapunov functions $V_1(\mathbf{z}), \ldots, V_N(\mathbf{z})$ are constructed which are continuously differentiable for $\mathbf{z} \in R^n$, positive definite and positive homogeneous of the degree $\gamma > 1$, and such that the functions $W_{ss}(\mathbf{z}) = \left(\frac{\partial V_s}{\partial \mathbf{z}}, \mathbf{F}^{(s)}(\mathbf{z})\right)$, $s = 1, \ldots, N$, are negative definite.

Remark 1. In [22,23] it was proved that fulfillment of Assumption 1 implies the existence of the required Lyapunov functions.

In view of the homogeneous functions properties [22], the estimates

$$a_{1s}\|\mathbf{z}\|^\gamma \le V_s(\mathbf{z}) \le a_{2s}\|\mathbf{z}\|^\gamma, \tag{5}$$

$$\left\|\frac{\partial V_s(\mathbf{z})}{\partial \mathbf{z}}\right\| \le a_{3s}\|\mathbf{z}\|^{\gamma-1}, \qquad W_{ss}(\mathbf{z}) \le b_{ss}\|\mathbf{z}\|^{\gamma+\mu-1}$$

hold for all $\mathbf{z} \in R^n$. Here

$$a_{1s} = \min_{\|\mathbf{z}\|=1} V_s(\mathbf{z}) > 0, \quad a_{2s} = \max_{\|\mathbf{z}\|=1} V_s(\mathbf{z}) > 0,$$

$$a_{3s} = \max_{\|\mathbf{z}\|=1}\left\|\frac{\partial V_s(\mathbf{z})}{\partial \mathbf{z}}\right\| \ge 0, \quad b_{ss} = \max_{\|\mathbf{z}\|=1} W_{ss}(\mathbf{z}) < 0$$

for $s = 1,\ldots,N$, and $\|\cdot\|$ denotes the Euclidean norm of a vector.

Consider the derivatives of the partial Lyapunov functions with respect to "foreign" subsystems

$$W_{sj}(\mathbf{z}) = \left(\frac{\partial V_s}{\partial \mathbf{z}}, \mathbf{F}^{(j)}(\mathbf{z})\right), \quad s \ne j; \ s,j = 1,\ldots,N.$$

Functions $W_{sj}(\mathbf{z})$ are continuous and positive homogeneous of the degree $\gamma + \mu - 1$, and for all $\mathbf{z} \in R^n$ the estimates

$$W_{sj}(\mathbf{z}) \le b_{sj}\|\mathbf{z}\|^{\gamma+\mu-1}, \quad s \ne j; \ s,j = 1,\ldots,N,$$

are valid, where $b_{sj} = \max_{\|\mathbf{z}\|=1} W_{sj}(\mathbf{z})$ [22].

It is worth mentioning, that for $s \ne j$, every number b_{sj} can be either non-negative or negative, since $V_s(\mathbf{z})$ can be Lyapunov function not only for "native" subsystem but also for "foreign" ones.

Applying the approaches proposed in [17,18], construct a common Lyapunov function for family (4) in the form

$$v(\mathbf{z}) = \sum_{s=1}^{N} \lambda_s V_s(\mathbf{z}). \tag{6}$$

Here $\lambda_1,\ldots,\lambda_N$ are positive coefficients. Function $v(\mathbf{z})$ is positive definite. In [18] it was proved that if $\lambda_1,\ldots,\lambda_N$ satisfy the inequalities

$$\sum_{s=1}^{N} b_{sj}\lambda_s < 0, \quad j = 1,\ldots,N, \tag{7}$$

then the derivative of $v(\mathbf{z})$ with respect to any subsystem from (2) is negative definite.

By the use of results of [21], it can be easily verified that then there exists a number $H > 0$ such that $\Delta v(\mathbf{z})\big|_{(s)} < 0$, $s = 1, \ldots, N$, for $0 < \|\mathbf{z}\| < H$, where $\Delta v(\mathbf{z})\big|_{(s)}$ denotes the difference of function (6) with respect to sth subsystem from (4). Hence, (6) is the common Lyapunov function for (4) satisfying the conditions of discrete counterpart of the Lyapunov asymptotic stability theorem [20].

Thus, the following theorem is valid.

Theorem 1. *Let Assumptions 1 and 2 be fulfilled. If the system of linear inequalities* (7) *admits a positive solution, then the zero solution of system* (3) *is uniformly asymptotically stable for any switching law.*

Remark 2. Computational procedures for finding positive solutions for systems of algebraic inequalities have been suggested in [19,24–27]. It is obvious that the worst situation for solvability of (7) is the one, where $b_{sj} \geq 0$ for $s \neq j$; $s, j = 1, \ldots, N$. In this case, (7) admits a positive solution if and only if the matrix $\mathbf{B} = (b_{sj})_{s,j=1}^{N}$ satisfies the Sevastyanov – Kotelyanskij conditions [28]: $(-1)^q \det (b_{sj})_{s,j=1}^{q} > 0$, $q = 1, \ldots, N$. These conditions form a stability criterion of linear comparison systems, and therefore they are commonly used for stability investigation of complex systems by means of vector Lyapunov functions [28].

4   Stability Analysis via Multiple Lyapunov Functions

Assume now that we failed to construct a common Lyapunov function for subsystems (4). In this case, to prove stability of a switched system, one should restrict the class of admissible switching signals [12,13]. These restrictions usually represent the lower bounds on the intervals between consecutive switching times. The general approach for finding such bounds is based on the use of multiple Lyapunov functions [12,15]. However, this approach has been developed mainly for families of subsystems with exponentially stable zero solutions. In [18], it was used for the stability analysis of switched homogeneous system (1). In the present section, we will extend this approach to the switched difference system (3).

Without loss of generality, we assume that the interval $(0, +\infty)$ contains the infinite number of switching instants. Let θ_i, $i = 1, 2, \ldots$, be the switching times, $0 < \theta_1 < \theta_2 < \ldots$, and $\theta_0 = 0$.

Consider the case where the switching instants θ_i, $i = 1, 2, \ldots$, are given, while the order of switching between subsystems may be unknown. Let Assumptions 1 and 2 be fulfilled.

In a similar way as in [21], it can be shown that there exist numbers $\alpha > 0$ and $H > 0$ such that the inequalities

$$\Delta V_s\big|_{(s)} \leq -\alpha V_s^{1+\frac{\mu-1}{\gamma}}(\mathbf{x}(k)), \quad s = 1, \ldots, N, \tag{8}$$

hold for $\|\mathbf{x}(k)\| < H$.

In [21] the following lemma has been proved.

Lemma 1. *If a sequence $\{v_k\}$ satisfies the condition*

$$0 \leq v_{k+1} \leq v_k - \alpha v_k^\lambda, \qquad k = 0, 1, \ldots,$$

with $\alpha > 0$, $\lambda > 1$, $v_0 \geq 0$, and $\alpha \lambda v_0^{\lambda-1} \leq 1$, then

$$v_k \leq v_0 \left(1 + \alpha(\lambda - 1)v_0^{\lambda-1}k\right)^{-\frac{1}{\lambda-1}}$$

for all $k = 0, 1, \ldots$

Application of this lemma to inequalities (8) allows to obtain the estimates for solutions of subsystems (4).

Denote

$$c = \max_{s,j=1,\ldots,N} \max_{\|\mathbf{z}\|=1} \frac{V_s(\mathbf{z})}{V_j(\mathbf{z})}, \qquad b = c^{-\frac{\mu-1}{\gamma}}.$$

Then $c \geq 1$, $0 < b \leq 1$, and

$$V_s(\mathbf{z}) \leq cV_j(\mathbf{z}), \qquad s, j = 1, \ldots, N, \tag{9}$$

for $\mathbf{z} \in R^n$.

Remark 3. In the case when $c = 1$, one gets $V_1(\mathbf{z}) = \ldots = V_N(\mathbf{z})$, i.e., for subsystems (4) there exists a common Lyapunov function, and the zero solution of the corresponding switched system is uniformly asymptotically stable for any switching law. In view of this fact, we assume subsequently that $c > 1$.

Let $T_i = \theta_i - \theta_{i-1}$, $i = 1, 2, \ldots$; $\psi(q, 1) = 0$, and $\psi(q, p) = \sum_{i=1}^{p-1} T_{q+i} b^{p-i}$ for $p = 2, 3, \ldots$, $q = 1, 2, \ldots$; and $\mathbf{x}(k, \mathbf{x}_0, k_0)$ denote the solution of (3) passing through $\mathbf{x}_0$ at $k = k_0$.

Theorem 2. *Let Assumptions 1 and 2 be fulfilled, and the switching times θ_i, $i = 1, 2, \ldots$, be given. If the condition*

$$\psi(q, p) \to +\infty \quad as \quad p \to +\infty \tag{10}$$

is valid for any positive integer q, then the zero solution of system (3) is asymptotically stable. In the case when the tendency (10) is uniform with respect to $q = 1, 2, \ldots$, the zero solution of system (3) is uniformly asymptotically stable.

Proof. By using partial Lyapunov functions $V_1(\mathbf{z}), \ldots, V_N(\mathbf{z})$, construct the multiple Lyapunov function $V_{\sigma(k)}(\mathbf{z})$ corresponding to the switching law $\sigma(k)$ [15].

Choose a number ε such that $0 < \varepsilon < H$, and

$$\alpha \left(1 + \frac{\mu - 1}{\gamma}\right) V_s^{\frac{\mu-1}{\gamma}}(\mathbf{z}) \leq 1, \qquad s = 1, \ldots, N, \tag{11}$$

for $\|\mathbf{z}\| < \varepsilon$.

Consider a solution $\mathbf{x}(k)$ of (3) starting at $k = k_0$ from the point $\mathbf{x}_0$, where $k_0 \geq 0$, $0 < \|\mathbf{x}_0\| < \varepsilon$. Find the positive integer q such that $k_0 \in [\theta_{q-1}, \theta_q)$.

Assume that the solution $\mathbf{x}(k)$ remains in the region $\|\mathbf{z}\| < \varepsilon$ for $k = k_0, \ldots, \tilde{k}$. Applying Lemma 1 to the $\sigma(\theta_{q-1})$th inequality from (8), we obtain that if $k_0 < \tilde{k} \leq \theta_q$, then

$$V_{\sigma(\theta_{q-1})}^{-\frac{\mu-1}{\gamma}}(\mathbf{x}(\tilde{k})) \geq V_{\sigma(\theta_{q-1})}^{-\frac{\mu-1}{\gamma}}(\mathbf{x}_0) + \frac{\alpha(\mu-1)}{\gamma}(\tilde{k} - k_0). \tag{12}$$

In the case when $\tilde{k} > \theta_q$, there exists the positive integer p satisfying the condition $\theta_{q+p-1} < \tilde{k} \leq \theta_{q+p}$. It should be mentioned that $p \to \infty$ as $\tilde{k} \to +\infty$. Applying successively Lemma 1 to the corresponding inequalities from (8) and taking into account estimates (9), we obtain

$$\begin{aligned}
V_{\sigma(\theta_{q+p-1})}^{-\frac{\mu-1}{\gamma}}(\mathbf{x}(\tilde{k})) &\geq V_{\sigma(\theta_{q+p-1})}^{-\frac{\mu-1}{\gamma}}(\mathbf{x}(\theta_{q+p-1})) + \frac{\alpha(\mu-1)}{\gamma}(\tilde{k} - \theta_{q+p-1}) \\
&\geq bV_{\sigma(\theta_{q+p-2})}^{-\frac{\mu-1}{\gamma}}(\mathbf{x}(\theta_{q+p-1})) + x\frac{\alpha(\mu-1)}{\gamma}(\tilde{k} - \theta_{q+p-1}) \geq \ldots \\
&\geq b^p V_{\sigma(\theta_{q-1})}^{-\frac{\mu-1}{\gamma}}(\mathbf{x}_0) + \frac{\alpha(\mu-1)}{\gamma}\left((\tilde{k} - \theta_{q+p-1}) + \psi(q,p) + b^p(\theta_q - k_0)\right).
\end{aligned} \tag{13}$$

With the aid of (5), (12) and (13), it is easy to show that

$$\|\mathbf{x}(\tilde{k})\| \leq \tilde{a}_1^{-\frac{1}{\gamma}}\left(\tilde{a}_2^{-\frac{\mu-1}{\gamma}}\|\mathbf{x}_0\|^{1-\mu} + \frac{\alpha(\mu-1)}{\gamma}(\tilde{k} - k_0)\right)^{-\frac{1}{\mu-1}} \tag{14}$$

for $\tilde{k} = k_0, \ldots, \theta_q$, and

$$\begin{aligned}
\|\mathbf{x}(\tilde{k})\| \leq \tilde{a}_1^{-\frac{1}{\gamma}}\Big(b^p \tilde{a}_2^{-\frac{\mu-1}{\gamma}}\|\mathbf{x}_0\|^{1-\mu} + \frac{\alpha(\mu-1)}{\gamma}\Big(\left(\tilde{k} - \theta_{q+p-1}\right) \\
+ \psi(q,p) + b^p\left(\theta_q - k_0\right)\Big)\Big)^{-\frac{1}{\mu-1}}
\end{aligned} \tag{15}$$

for $\tilde{k} = \theta_{q+p-1} + 1, \ldots, \theta_{q+p}$; $p \geq 1$. Here

$$\tilde{a}_1 = \min\{a_{11}, \ldots, a_{1N}\}, \qquad \tilde{a}_2 = \max\{a_{21}, \ldots, a_{2N}\}.$$

Let relation (10) be valid. For given values of ε and q, choose a positive integer number p_0 such that

$$\psi(q,p) > \frac{\gamma}{\alpha(\mu-1)}\left(\varepsilon \tilde{a}_1^{\frac{1}{\gamma}}\right)^{1-\mu} \qquad \text{for} \quad p \geq p_0.$$

Taking

$$\delta = \varepsilon\,(\tilde{a}_1/\tilde{a}_2)^{1/\gamma}\, b^{p_0/(\mu-1)}, \tag{16}$$

and using estimates (14) and (15), one gets that if for a solution $\mathbf{x}(k)$ of system (3) the inequality $\|\mathbf{x}(k_0)\| < \delta$ is valid, then $\|\mathbf{x}(k)\| < \varepsilon$ for all $k \geq k_0$. Moreover, from (15) it follows that $\|\mathbf{x}(k)\| \to 0$ as $k \to +\infty$. Thus, the zero solution of (3) is asymptotically stable.

Assume next that (10) holds uniformly with respect to $q = 1, 2, \ldots$. Let us prove the uniform asymptotic stability.

In this case, the number p_0 can be chosen independently of q. Hence, the zero solution is uniformly stable.

Fix a number $\varepsilon \in (0, H)$ such that the inequality (11) is valid for $\|\mathbf{z}\| < \varepsilon$, and define the number δ by the formula (16). To prove the uniform attraction, it should be shown that for any $\varepsilon' > 0$ there exists a number $T > 0$ such that if $k_0 \geq 0$, $\|\mathbf{x}_0\| < \delta$, then $\|\mathbf{x}(k, \mathbf{x}_0, k_0)\| < \varepsilon'$ for all $k \geq k_0 + T$.

For given $\varepsilon' > 0$, find $\delta' > 0$ according to the definition of uniform stability. We obtain that if a solution $\mathbf{x}(k)$ of (3) enters into the region $\|\mathbf{z}\| < \delta'$ at $k = k_1 \geq 0$, then $\|\mathbf{x}(k)\| < \varepsilon'$ for $k \geq k_1$.

Choose $p' \geq 1$, such that

$$\psi(q, p) > \frac{\gamma}{\alpha(\mu - 1)} \left(\delta' \tilde{a}_1^{\frac{1}{\gamma}} \right)^{1-\mu}$$

for $p \geq p'$ and for all $q = 1, 2, \ldots$. Let

$$T' = \frac{\gamma}{\alpha(\mu - 1)} b^{-p'} \left(\delta' \tilde{a}_1^{\frac{1}{\gamma}} \right)^{1-\mu}.$$

Consider a solution $\mathbf{x}(k, \mathbf{x}_0, k_0)$ of (3) with initial conditions satisfying the inequalities $k_0 \geq 0$, $0 < \|\mathbf{x}_0\| < \delta$. If the number of switching times in the interval $[k_0, k_0 + T']$ exceeds p', then

$$\|\mathbf{x}(k_0 + [T'] + 1, \mathbf{x}_0, k_0)\| \leq \tilde{a}_1^{-\frac{1}{\gamma}} \left(\frac{\alpha(\mu - 1)}{\gamma} \psi(q, p) \right)^{-\frac{1}{(\mu-1)}} < \delta',$$

where $[T']$ denotes the integral part of T'.

Otherwise, we obtain

$$\|\mathbf{x}(k_0 + [T'] + 1, \mathbf{x}_0, k_0)\| \leq \tilde{a}_1^{-\frac{1}{\gamma}} \left(\frac{\alpha(\mu - 1)}{\gamma} b^{p'} T' \right)^{-\frac{1}{(\mu-1)}} < \delta'.$$

Hence, $\|\mathbf{x}(k, \mathbf{x}_0, k_0)\| < \varepsilon'$ for $k \geq k_0 + [T'] + 1$. This completes the proof.

Corollary 1. *Let Assumptions 1 and 2 be fulfilled, and the switching times* θ_i, $i = 1, 2, \ldots$, *be given. If* $T_i \to +\infty$ *as* $i \to \infty$, *then the zero solution of system* (3) *is uniformly asymptotically stable.*

Remark 4. It is a fairly well-known fact that for any family consisting of a finite number of linear time invariant asymptotically stable difference subsystems there exists a number $L > 0$ (dwell-time), such that the corresponding switched system is also asymptotically stable providing that the intervals between consecutive switching times are not less than L [12, 13]. Theorem 2 does not permit to obtain the similar result for the family of nonlinear subsystems (4). If $T_i = L = \text{const} > 0$, $i = 1, 2, \ldots$, then condition (10) is not fulfilled for any choice of L. However, for nonlinear switched system (3), a positive lower bound for the values of $T_1, T_2, \ldots$ can be found guaranteeing the practical stability [29] of the zero solution.

Corollary 2. *Let Assumptions 1 and 2 be fulfilled, and the switching times θ_i, $i = 1, 2, \ldots$, be given. Then there exists a positive number δ such that for any $\varepsilon > 0$ one can choose $L_1 > 0$ and $L_2 > 0$ satisfying the following condition: if $T_i \geq L_1$, $i = 1, 2, \ldots$, and for a solution $\mathbf{x}(k, \mathbf{x}_0, k_0)$ of (3) the inequalities $k_0 \geq 0$, $\|\mathbf{x}_0\| < \delta$ are valid, then $\|\mathbf{x}(k, \mathbf{x}_0, k_0)\| < \varepsilon$ for $k \geq k_0 + L_2$.*

5   Asymptotic Stability Conditions in the Case Of Complete Information on the Switching Law

Assume now that we know not only the switching instants θ_i, $i = 1, 2, \ldots$, but also the order of switching between the subsystems. In this case, another approach for the stability analysis can be used [16]. Choose a subsystem from family (4) and determine relationship between this subsystem activity intervals and those of the remaining subsystems under which it is possible to guarantee the asymptotic stability of the zero solution of switched system (3).

We replace Assumptions 1 and 2 by weaker ones formulated (for definiteness) for the choice of the first subsystem.

Assumption 3. The zero solution of the first subsystem from family (2) is asymptotically stable.

Assumption 4. For the first subsystem from family (2), a Lyapunov function $V_1(\mathbf{z})$ is constructed which is continuously differentiable for $\mathbf{z} \in R^n$, positive definite and positive homogeneous of the degree $\gamma > 1$, and such that the function $W_{11}(\mathbf{z}) = \left(\frac{\partial V_1}{\partial \mathbf{z}}, \mathbf{F}^{(1)}(\mathbf{z})\right)$ is negative definite.

Under Assumptions 3 and 4, we obtain

$$a_{11}\|\mathbf{z}\|^\gamma \leq V_1(\mathbf{z}) \leq a_{21}\|\mathbf{z}\|^\gamma, \tag{17}$$

where

$$a_{11} = \min_{\|\mathbf{z}\|=1} V_1(\mathbf{z}) > 0, \quad a_{21} = \max_{\|\mathbf{z}\|=1} V_1(\mathbf{z}) > 0.$$

In a similar way as in [21], it is easy to show the existence of a number $H > 0$ such that the inequalities

$$\Delta V_1\big|_{(s)} \leq \beta_s V_1^{1 + \frac{\mu - 1}{\gamma}}(\mathbf{x}(k)), \quad s = 1, \ldots, N, \tag{18}$$

are valid for $\|\mathbf{x}(k)\| < H$. Here $\beta_1, \ldots, \beta_N$ are constants with $\beta_1 < 0$.

Lemma 2. *If a sequence $\{v_k\}$ satisfies the condition*

$$0 \leq v_{k+1} \leq v_k + \alpha v_k^\lambda, \qquad k = 0, 1, \ldots,$$

with $\alpha > 0$, $\lambda > 1$, $v_0 \geq 0$. Then for all $k \geq 0$ such that $\alpha(\lambda - 1)v_0^{\lambda-1}k < 1$ the estimate

$$v_k \leq v_0 \left(1 - \alpha(\lambda - 1)v_0^{\lambda-1}k\right)^{-\frac{1}{\lambda-1}} \tag{19}$$

holds.

Proof. We proceed by induction. Inequality (19) is true for $k = 0$. Assume that it is valid for $k = 0, \ldots, l$, and

$$\alpha(\lambda - 1)v_0^{\lambda-1}(l+1) < 1. \tag{20}$$

Then

$$v_{l+1} \leq v_l + \alpha v_l^\lambda \leq v_0 \left(1 - \alpha(\lambda - 1)v_0^{\lambda-1}l\right)^{-\frac{1}{\lambda-1}}$$

$$+ \alpha v_0^\lambda \left(1 - \alpha(\lambda - 1)v_0^{\lambda-1}l\right)^{-\frac{\lambda}{\lambda-1}} = v_0 \left(1 - \alpha(\lambda - 1)v_0^{\lambda-1}(l+1)\right)^{-\frac{1}{\lambda-1}} \varphi(y),$$

where

$$\varphi(y) = \left(1 + \frac{y}{\lambda - 1}\right)(1 - y)^{\frac{1}{\lambda-1}}, \qquad y = \frac{\alpha(\lambda - 1)v_0^{\lambda-1}}{1 - \alpha(\lambda - 1)v_0^{\lambda-1}l}.$$

Under the condition (20), we obtain $0 \leq y < 1$. The function $\varphi(y)$ is decreasing on the interval $[0, 1]$, and $\varphi(0) = 1$. Hence, (19) holds for $k = l + 1$ as well. This completes the proof.

For given switching law $\sigma(k)$, construct the auxiliary functions $\eta(k) = -\beta_{\sigma(k)}$, $k = 0, 1, \ldots$; $\chi(k_0, k_0) = 0$, and $\chi(k_0, k) = \sum\limits_{i=k_0}^{k-1} \eta(i)$ for $k > k_0$, $k_0 = 0, 1, \ldots$.

Theorem 3. *Let Assumptions 3 and 4 be fulfilled, and the switching law $\sigma(k)$ be given. If the condition*

$$\chi(k_0, k) \to +\infty \quad as \quad k \to +\infty \tag{21}$$

is valid for any $k_0 \geq 0$, then the zero solution of system (3) is asymptotically stable. In the case when the tendency (21) is uniform with respect to $k_0 = 0, 1, \ldots$, the zero solution of system (3) is uniformly asymptotically stable.

Proof. Consider a switching law $\sigma(k)$. Construct the corresponding function $\eta(k)$. Let $k_0 \geq 0$ be given. Choose a number ε such that $0 < \varepsilon < H$, and

$$-\beta_1 \left(1 + \frac{\mu - 1}{\gamma}\right) V_1^{\frac{\mu-1}{\gamma}}(\mathbf{z}) \leq 1$$

for $\|\mathbf{z}\| < \varepsilon$.

Consider a solution $\mathbf{x}(k)$ of (3) starting at $k = k_0$ from the point $\mathbf{x}_0$, where $k_0 \geq 0$, $0 < \|\mathbf{x}_0\| < \varepsilon$. Applying Lemmas 1 and 2 to the inequalities (18), we obtain that if the solution $\mathbf{x}(k)$ remains in the region $\|\mathbf{z}\| < \varepsilon$ for $k = k_0, \ldots, \tilde{k}$, and the inequality $V_1^{-\frac{\mu-1}{\gamma}}(\mathbf{x}_0) + \frac{(\mu-1)}{\gamma}\chi(k_0, k) > 0$ holds, then

$$V_1^{-\frac{\mu-1}{\gamma}}(\mathbf{x}(k)) \geq V_1^{-\frac{\mu-1}{\gamma}}(\mathbf{x}_0) + \frac{(\mu-1)}{\gamma}\chi(k_0, k)$$

for $k = k_0, \ldots, \tilde{k}$.

By using estimates (17), one gets

$$\|\mathbf{x}(k)\| \leq a_{11}^{-\frac{1}{\gamma}}\left(a_{21}^{-\frac{\mu-1}{\gamma}}\|\mathbf{x}_0\|^{1-\mu} + \frac{(\mu-1)}{\gamma}\chi(k_0, k)\right)^{-\frac{1}{\mu-1}}, \quad k = k_0, \ldots, \tilde{k}. \quad (22)$$

If (21) is fulfilled, then there exists a constant ρ_0 such that $\chi(k_0, k) \geq \rho_0$ for all $k \geq k_0$. Choose a number $\delta > 0$ satisfying the condition

$$a_{21}^{-\frac{\mu-1}{\gamma}}\delta^{1-\mu} + \frac{(\mu-1)\rho_0}{\gamma} > \left(a_{11}^{\frac{1}{\gamma}}\varepsilon\right)^{1-\mu}.$$

From the estimate (22) it follows that if $0 < \|\mathbf{x}_0\| < \delta$, then $\|\mathbf{x}(k)\| < \varepsilon$ for all $k \geq k_0$, and furthermore $\|\mathbf{x}(k)\| \to 0$ as $k \to +\infty$. Thus, the zero solution of system (3) is asymptotically stable.

In the case when (21) holds uniformly with respect to $k_0 = 1, 2, \ldots$, the numbers ρ_0 and δ can be chosen independently of k_0. Hence, the zero solution is uniformly stable. By the use of estimate (22), it can be shown that in the case considered, the uniform attraction takes place as well. This completes the proof.

Remark 5. Theorem 3 remains valid also in the case when the zero solutions of a part of subsystems numbered $2, \ldots, N$ or all of these subsystems are not asymptotically stable.

Let us show next that, unlike Theorems 2 and 3 can be used for the finding of a fixed lower bound on the intervals between consecutive switching times guaranteeing asymptotic stability of the zero solution of the corresponding switched system.

Consider again the case when Assumptions 3 and 4 are fulfilled. Let $V_1(\mathbf{z})$ be the Lyapunov function not only for the first subsystem from the family (4) but also for a part of remaining subsystems. Then, in the inequalities (18), constants β_s corresponding to such subsystems are negative.

Assumption 5. Let $\beta_s < 0$ for $s = 1, \ldots, l$, and $\beta_s \geq 0$ for $s = l+1, \ldots, N$, where $1 \leq l < N$.

Denote by A the set of subsystems of the family (4) numbered $1, \ldots, l$, while by B the set of the remained subsystems. For a given switched law $\sigma(k)$, construct the subsequence $\{\tilde{\theta}_i\}_{i=1}^{\infty} \subset \{\theta_i\}_{i=1}^{\infty}$ consisting of all the switching instants in which switching occurs between a subsystem from the one set to a subsystem from the other.

Without loss of generality, we assume that for $k = 0$ a subsystem from the set A is active. Let $\tilde{\theta}_0 = 0$. Then for $k = \tilde{\theta}_{2i-2}, \ldots, \tilde{\theta}_{2i-1} - 1$ switching occurs only between subsystems from the set A, while for $k = \tilde{\theta}_{2i-1}, \ldots, \tilde{\theta}_{2i} - 1$ only between subsystems from the set B, $i = 1, 2, \ldots$. Denote $\tilde{T}_i = \tilde{\theta}_i - \tilde{\theta}_{i-1}$, $i = 1, 2, \ldots$.

Theorem 4. *Let Assumptions 3, 4 and 5 be fulfilled, and the switching law $\sigma(k)$ be given. Then, for any $L_2 > 0$, there exists a number $L_1 > 0$ such that if $\tilde{T}_{2i-1} \geq L_1$ and $\tilde{T}_{2i} \leq L_2$ for all $i = 1, 2, \ldots$, then the zero solution of system (3) is uniformly asymptotically stable.*

Proof. Choose an arbitrary number $L_2 > 0$ and find $L_1 > 0$ such that $\alpha_1 L_1 + \alpha_2 L_2 < 0$. Here $\alpha_1 = \max_{s=1,\ldots,l}\beta_s$, $\alpha_2 = \max_{s=l+1,\ldots,N}\beta_s$. Then the condition (21) is fulfilled uniformly with respect to $k_0 = 0, 1, \ldots$. Applying Theorem 3, we obtain that the zero solution of system (3) is uniformly asymptotically stable. This completes the proof.

6 Example

Let family (4) consists of two subsystems

$$x_1(k+1) = x_1(k) - x_1^3(k) - x_2^3(k), \quad x_2(k+1) = x_2(k) + 10x_1^3(k) - x_2^3(k), \quad (23)$$

$$x_1(k+1) = x_1(k) - x_1^3(k) - 10x_2^3(k), \quad x_2(k+1) = x_2(k) + x_1^3(k) - x_2^3(k). \quad (24)$$

Consider the corresponding homogeneous differential systems

$$\dot{z}_1 = -z_1^3 - z_2^3, \quad \dot{z}_2 = 10z_1^3 - z_2^3, \tag{25}$$

$$\dot{z}_1 = -z_1^3 - 10z_2^3, \quad \dot{z}_2 = z_1^3 - z_2^3. \tag{26}$$

Construct homogeneous Lyapunov functions for (25) and (26) in the form

$$V_1(z_1, z_2) = 10z_1^4 + z_2^4, \quad V_2(z_1, z_2) = z_1^4 + 10z_2^4.$$

Thus, in this case $n = 2$, $N = 2$, $\mu = 3$, $\gamma = 4$.

Differentiating $V_1(z_1, z_2)$ and $V_2(z_1, z_2)$ with respect to systems (25) and (26), we obtain

$$\dot{V}_1\big|_{(25)} = -40z_1^6 - 4z_2^6, \quad \dot{V}_1\big|_{(26)} = -40z_1^6 - 396z_1^3 z_2^3 - 4z_2^6,$$

$$\dot{V}_2\big|_{(25)} = -4z_1^6 + 396z_1^3 z_2^3 - 40z_2^6, \quad \dot{V}_2\big|_{(26)} = -4z_1^6 - 40z_2^6.$$

Using the numerical analysis, it can be shown that $b_{11} = b_{22} \approx -2.3$, $b_{12} = b_{21} \approx 44.6$.

Applying Theorem 1, consider the inequalities system

$$-2.3\lambda_1 + 44.6\lambda_2 < 0, \quad 44.6\lambda_1 - 2.3\lambda_2 < 0.$$

It is easily verified that this system does not have a positive solution. Therefore, we failed to construct a common Lyapunov function for subsystems (23), (24) on the basis of the approach proposed in Sect. 3.

Now apply the results of Sect. 4. In this case $c = 10$, $b = 1/\sqrt{10}$. By the use of Theorem 2, we obtain that if the condition

$$\sum_{i=1}^{p-1} T_{i+1} 10^{\frac{i-p}{2}} \to +\infty \quad \text{as} \quad p \to \infty$$

is met, then, for any order of switching between the subsystems (23) and (24), the zero solution of the corresponding switched difference system is asymptotically stable.

Next, assume that we know not only the switching instants θ_i, $i = 1, 2, \ldots$, but also the order of switching between the subsystems. Without loss of generality, we consider the case when on the intervals $[\theta_{2i-2}, \theta_{2i-1})$ the subsystem (23) is active, while on the intervals $[\theta_{2i-1}, \theta_{2i})$ subsystem (24) is, $i = 1, 2, \ldots$.

For the function $V_1(z_1, z_2)$ the estimates

$$\dot{V}_1\big|_{(25)} \leq -1.2 \cdot V_1^{\frac{3}{2}}, \qquad \dot{V}_1\big|_{(26)} \leq 23.1 \cdot V_1^{\frac{3}{2}}.$$

are valid. Hence, for any $\beta_1 \in (-1.2, 0)$ and $\beta_2 > 23.1$, there exists a number $H > 0$ such that

$$\Delta V_1\big|_{(23)} \leq \beta_1 V_1^{\frac{3}{2}}, \qquad \Delta V_1\big|_{(24)} \leq \beta_2 V_1^{\frac{3}{2}}$$

for $\|\mathbf{x}(k)\| < H$. By the use of Theorem 4, we obtain that if the inequalities $1.2 \cdot T_{2i-1} \geq 23.1 \cdot T_{2i} + \delta$, $i = 1, 2, \ldots$, hold, where δ is a positive constant, then the zero solution of the switched difference system generated by the family (23), (24) is asymptotically stable.

7   Conclusion

In this paper, a class of nonlinear switched difference systems is studied. Systems of the class appear as computational schemes for switched differential systems with homogeneous right-hand sides. By the use of the Lyapunov direct method, some sufficient conditions of the asymptotic stability of solutions for difference systems are obtained. The fulfilment of these conditions provides the dynamical consistency between the discrete equations considered and their continuous analogues and consequently gives reliable numerical results. It is worth mentioning that although the stability conditions are independent of the digitization step h, the region of the asymptotic stability of the zero solution depends on the chosen value of h.

In this paper, we studied the difference systems that appear when the Euler numerical integration scheme is used. However, the approaches proposed are applicable to the difference systems arising from numerical integration based on the Runge-Kutta or Adams schemes as well. As we here have considered the notion of homogeneity with respect to the standard dilation, future research may find it valuable to extend the obtained results to the case of systems with right-hand sides being homogeneous functions with respect to an arbitrary dilation.

Another direction for further research is application of the proposed approaches to problems of mechanics and electrodynamics.

Acknowledgments. This work is supported by the St. Petersburg State University (project no. 9.38.674.2013), the Russian Foundation for Basic Research (grant no. 15-58-53017), NSF of China (project no. 61273006), and High Technology Research and Development Program of China (863 Program) (project no. 2011AA110301).

References

1. Dekker, K., Verwer, J.G.: Stability of Runge-Kutta Methods for Stiff Nonlinear Differential Equations. North-Holland, Amsterdam, New York, Oxford (1984)
2. Mickens, R.E.: Applications of Nonstandard Finite Difference Schemes. World Scientific, Singapore (2000)
3. Roeger, L.-I.W.: Local stability of Euler's and Kahan's methods. J. Differ. Eqn. Appl. **10**(6), 601–614 (2004)
4. Mickens, R.E.: A Note on a Discretization Scheme for Volterra Integro-differential Equations that Preserves Stability and Boundedness. J. Differ. Eqn. Appl. **13**(6), 547–550 (2007)
5. Volkova, A.S., Gnilitskaya, Y.A., Provotorov, V.V.: On the solvability of boundary-value problems for parabolic and hyperbolic equations on geometrical graphs. Autom. Remote Control **75**(2), 405–412 (2014)
6. Aleksandrov, A.Y., Chen, Y., Platonov, A.V., Zhang, L.: Stability analysis and uniform ultimate boundedness control synthesis for a class of nonlinear switched difference systems. J. Differ. Eqn. Appl. **18**(9), 1545–1561 (2012)
7. Sanz-Serna, J.M.: Symplectic integrators for hamiltonian problems: an overview. Acta Numer. **1**, 243–286 (1992)
8. Zubov, V.I.: Conservative numerical methods for integrating differential equations in nonlinear mechanics. Doklady Math. **55**(3), 388–390 (1997)
9. Aleksandrov, A.Y., Zhabko, A.P.: Preservation of stability under discretization of systems of ordinary differential equations. Siberian Math. J. **51**(3), 383–395 (2010)
10. Kaufmann, E.R., Raffoul, Y.N.: Discretization scheme in volterra integro-differential equations that preserves stability and boundedness. J. Differ. Eqn. Appl. **12**(7), 731–740 (2006)
11. Patidar, K.G.: On the use of nonstandard finite difference methods. J. Differ. Eqn. Appl. **11**(8), 735–758 (2005)
12. Liberzon, D., Morse, A.S.: Basic problems in stability and design of switched systems. IEEE Control Syst. Mag. **19**(15), 59–70 (1999)
13. Shorten, R., Wirth, F., Mason, O., Wulf, K., King, C.: Stability criteria for switched and hybrid systems. SIAM Rev. **49**(4), 545–592 (2007)
14. Decarlo, R.A., Branicky, M.S., Pettersson, S., Lennartson, B.: Perspectives and results on the stability and stabilizability of hybrid systems. Proc. IEEE **88**(7), 1069–1082 (2000)
15. Branicky, M.S.: Multiple lyapunov functions and other analysis tools for switched and hybrid systems. IEEE Trans. Autom. Control **43**(4), 475–482 (1998)
16. Michel, A.N., Hou, L.: Stability results involving time-averaged lyapunov function derivatives. Nonlinear Anal. Hybrid Syst. **3**, 51–64 (2009)
17. Bobylev, N.A., Il'in, A.V., Korovin, S.K., Fomichev, V.V.: On the stability of families of dynamical systems. Diff. Eqn. **38**(4), 464–470 (2002)

18. Aleksandrov, A.Y., Kosov, A.A., Platonov, A.V.: On the asymptotic stability of switched homogeneous systems. Syst. Control Lett. **61**(1), 127–133 (2012)
19. Aleksandrov, A.Y., Platonov, A.V.: On absolute stability of one class of nonlinear switched systems. Autom. Remote Control **69**(7), 1101–1116 (2008)
20. Halanay, A., Wexler, D.: The Qualitative Theory of Systems with Impulse. Mir, Moscow (1971). (in Russian)
21. Aleksandrov, A., Zhabko, A.P.: Stability of solutions of nonlinear difference systems. Russ. Math. (Izvestiya VUZ. Matematika) **49**(2), 1–10 (2005)
22. Zubov, V.I.: Mathematical Methods for the Study of Automatical Control Systems. Pergamon Press, Jerusalem Academy Press, Oxford, Jerusalem (1962)
23. Rosier, L.: Homogeneous lyapunov function for homogeneous continuous vector field. Syst. Control Lett. **19**(6), 467–473 (1992)
24. Tschernikow, S.N.: Lineare Ungleichungen. VEB Deutscher Verlag der Wissenschaften, Berlin (1971)
25. Cox, D.A., Little, J., O'Shea, D.: Ideals, Varieties and Algorithms. Springer, New York (2007)
26. Uteshev, A.Y., Yashina, M.V.: Distance computation from an ellipsoid to a linear or a quadric surface in IR^n. In: Ganzha, V.G., Mayr, E.W., Vorozhtsov, E.V. (eds.) CASC 2007. LNCS, vol. 4770, pp. 392–401. Springer, Heidelberg (2007)
27. Platonov, A.V.: On Stability of Complex Nonlinear Systems. Izvestiya Akademii Nauk. Teoriya i Sistemy Upravleniya. (4), 41–46.(2004)
28. Abdullin, R.Z., Anapolsky, L., et al.: Vector Lyapunov Functions in Stability Theory. Advanced Series in Mathematical Science and Engineering. World Federation Publishers Company, Atlanta (1996)
29. La Salle, J., Lefschetz, S.: Stability by Lyapunov's Direct Method. Academic Press, New York, London (1961)

On Maxwell's Conjecture for Coulomb Potential Generated by Point Charges

Alexei Yu. Uteshev$^{(\boxtimes)}$ and Marina V. Yashina

Faculty of Applied Mathematics, St. Petersburg State University,
Universitetskij pr. 35, Petrodvorets, 198504 St. Petersburg, Russia
alexeiuteshev@gmail.com, marina.yashina@gmail.com

Abstract. The problem discussed herein is the one of finding the set of stationary points for the Coulomb potential function $F(P) = \sum_{j=1}^{K} m_j/|PP_j|$ for the cases of $K = 3$ and $K = 4$ positive charges $\{m_j\}_{j=1}^{K}$ fixed at the positions $\{P_j\}_{j=1}^{K} \subset \mathbb{R}^2$. Our approach is based on reducing the problem to that of evaluation of the number of real solution of an appropriate algebraic system of equations. We also investigate the bifurcation picture in the parameter domains.

Keywords: Coulomb potential · Stationary points · Maxwell's conjecture

1 Introduction

Given the coordinates of $K \geq 3$ points $\{P_j\}_{j=1}^{K} \subset \mathbb{R}^3$, find the coordinates of stationary points for the function

$$F(P) = \sum_{j=1}^{K} \frac{m_j}{|PP_j|} \cdot \tag{1}$$

Here $\{m_j\}_{j=1}^{K}$ are assumed to be real non-zero numbers and $|\cdot|$ stands for the Euclidian distance.

This problem can be viewed as a classical electrostatics one with the function (1) representing the Coulomb potential of the charges $\{m_j\}_{j=1}^{K}$ placed at fixed (stationary) positions $\{P_j\}_{j=1}^{K}$ in the space. Thus, the stated problem can be considered as an origin for the general problem of simulation or motions of charged particles in electric or magnetic fields; the stationary point of the potential corresponds then to the equilibrium position of a probe particle. On the other hand, the function (1) can also be interpreted as the Newton (gravitational) potential with $\{m_j\}_{j=1}^{K}$ treated as masses fixed at $\{P_j\}_{j=1}^{K}$. Despite its classical looking formulation, the problem has not been given a systematic exploration — with the exception of some special configurations like the one treated in [3] where the points $\{P_j\}_{j=1}^{K}$ make an equilateral polygon and all the charges $\{m_j\}_{j=1}^{K}$ are assumed to be equal. The difficulty of the problem can be acknowledged also from the state of the art with its part known as

© Springer-Verlag Berlin Heidelberg 2016
M.L. Gavrilova and C.J. Kenneth Tan (Eds.): Trans. on Comput. Sci. XXVII, LNCS 9570, pp. 68–80, 2016.
DOI: 10.1007/978-3-662-50412-3_5

Maxwell's Conjecture [7]. The total number of stationary points of any configuration with K charges in $\mathbb{R}^3$ never exceeds $(K-1)^2$.

This conjecture was investigated in [4, 8] with the aid of some topological principles. However, it remains still open even for the case of $K = 3$ equal charges.

The coordinates of stationary points of the function (1) satisfy the system of equations

$$\frac{DF}{DP} = \mathbb{O} \quad \Longleftrightarrow \quad \sum_{j=1}^{K} \frac{m_j(P - P_j)}{|PP_j|^3} = \mathbb{O}. \tag{2}$$

Solving this system with the aid of numerical iteration methods, like the gradient descent one, can lead one to poor convergence due to the unbounded growth of iteration values when the solution being searched lies in a close neighborhood of a charge location.

The present paper is devoted to an alternative approach for solving the system (2) based on symbolic computations. We first intend *to eliminate radicals* from the system (2), i.e. to reduce it to a system polynomially dependent on the coordinates of the point P. This can be done in different ways, and therefore it is quite reasonable to look for the procedure which can diminish the degrees of the final algebraic equations. For this purpose, we are going to exploit an approach suggested in the paper [11] where the general problem of finding the stationary point set for the function $F(P) = \sum_{j=1}^{K} m_j |PP_j|^L$ for arbitrary values of the exponent $L \neq 0$ was treated. Then, for the obtained algebraic system, we solve the problem of *localization of* its *solutions*, i.e. we aim at finding the true number of real solutions and separating them. The mathematical background for this approach is based on the technique of *elimination of variables* from the algebraic systems with the aid of *the resultant* computation. Our analysis of the behavior of the stationary point set in its dependency on the parameters involved into the problem (like the values of charges or coordinates of charge location) will be essentially based on the *discriminant* sign evaluation. We refer the reader to [2, 5, 10] for brushing up some basic results of Elimination Theory utilized in the foregoing sections.

We will deal here only with the cases of potential generated by 3 or 4 positive charges located <u>on the plane</u>. One possible misunderstanding should be cleared out in connection with this assumption. In some examples given below, we use the expressions *minimum* or *stable* stationary point. The *Earnshaw's theorem* states that a collection of point charges in $\mathbb{R}^3$ cannot be maintained in a stable stationary equilibrium configuration solely by the electrostatic interaction of the charges [9]. Therefore, the term *stability* hereinafter should be understood as the *conditional stability* in the plane of charges location.

2   Three Points

Let the points $\{P_j = (x_j, y_j)\}_{j=1}^3$ be noncollinear, i.e. the determinant

$$S = \begin{vmatrix} 1 & 1 & 1 \\ x_1 & x_2 & x_3 \\ y_1 & y_2 & y_3 \end{vmatrix} \tag{3}$$

does not vanish. For definiteness, we will assume hereinafter that the points P_1, P_2, P_3 are counted counterclockwise, i.e. the determinant (3) is positive.

Stationary points of the Coulomb potential

$$F(P) = \frac{m_1}{|PP_1|} + \frac{m_2}{|PP_2|} + \frac{m_3}{|PP_3|} \qquad (4)$$

are given by the system of Eq. (2) which, for this particular case, can be written down as

$$\begin{cases} \dfrac{m_1(x-x_1)}{|PP_1|^3} + \dfrac{m_2(x-x_2)}{|PP_2|^3} + \dfrac{m_3(x-x_3)}{|PP_3|^3} = 0, \\ \dfrac{m_1(y-y_1)}{|PP_1|^3} + \dfrac{m_2(y-y_2)}{|PP_2|^3} + \dfrac{m_3(y-y_3)}{|PP_3|^3} = 0. \end{cases} \qquad (5)$$

In order to transform this system into an algebraic one, the straightforward approach can be utilized consisting in successive squaring of the equations and eliminating the radicals one by one. If one denotes by A_1, A_2 and A_3 the summands in any of the above equations, then this procedure is executed as follows

$$A_1 + A_2 + A_3 = 0 \quad \Rightarrow \quad (A_1 + A_2)^2 = A_3^2 \quad \Rightarrow \quad (2\,A_1 A_2)^2 = (A_3^2 - A_1^2 - A_2^2)^2 \ .$$

However this approach (tackled in [6] for finding a boundary for the number of stationary points) results in a drastic increase of the order and complexity of the final algebraic equations. The resulting system can be reduced to an algebraic one

$$F_1(x, y, m_1, m_2, m_3) = 0, \quad F_2(x, y, m_1, m_2, m_3) = 0$$

where F_1, F_2 are polynomials of the degree 28 with respect to the variables x and y (and with the coefficients of the orders up to 10^{19} for Example 1 treated below). Finding all the real solutions of this system with the aid of elimination of variable procedure (resultant or the Gröbner basis computation [2]) is a hardly feasible task.

An alternative approach for reducing the system (5) to an algebraic one was suggested in [11]. It is based on the following result:

Theorem 1. *Set*

$$S_1(x, y) = \begin{vmatrix} 1 & 1 & 1 \\ x & x_2 & x_3 \\ y & y_2 & y_3 \end{vmatrix}, \quad S_2(x, y) = \begin{vmatrix} 1 & 1 & 1 \\ x_1 & x & x_3 \\ y_1 & y & y_3 \end{vmatrix}, \quad S_3(x, y) = \begin{vmatrix} 1 & 1 & 1 \\ x_1 & x_2 & x \\ y_1 & y_2 & y \end{vmatrix} \ . \qquad (6)$$

Any solution of the system (5) is a solution of the system

$$m_1 : m_2 : m_3 = |PP_1|^3 S_1(x, y) : |PP_2|^3 S_2(x, y) : |PP_3|^3 S_3(x, y) \ . \qquad (7)$$

The underlying idea of the proof of this theorem is simple: system (5) can be treated as a linear one with respect to the parameters m_1, m_2, m_3 and therefore can be resolved with the aid of Cramer's formulae, which are equivalent to (7).

Squaring the ratio (7) gives rise to the following algebraic system

$$m_2^2 S_1^2 |PP_1|^6 - m_1^2 S_2^2 |PP_2|^6 = 0, \quad m_2^2 S_3^2 |PP_3|^6 - m_3^2 S_2^2 |PP_2|^6 = 0 \ . \qquad (8)$$

Example 1. Let $P_1 = (1,1), P_2 = (5,1), P_3 = (2,6)$. Analyse the structure of the set of stationary points of the potential (4) for $m_1 = 1$ and for m_2, m_3 treated as parameters.

Solution. The system (8) is as follows

$$\begin{cases} \tilde{F}_1(x,y,m_2,m_3) = (5x+3y-28)^2(x^2+y^2-2x-2y+2)^3 m_2^2 \\ \qquad -(5x-y-4)^2(x^2+y^2-10x-2y+26)^3 = 0, \\ \tilde{F}_2(x,y,m_2,m_3) = (4y-4)^2(x^2+y^2-4x-12y+40)^3 m_2^2 \\ \qquad -m_3^2(5x-y-4)^2(x^2+y^2-10x-2y+26)^3 = 0 \end{cases} \quad (9)$$

and the degree of $\tilde{F}_1$ and $\tilde{F}_2$ with respect to the variables x and y equals 8. This time, in comparison with the approach mentioned at the beginning of the present section, it is realistic to eliminate any variable from this system. For instance, the resultant of these polynomials treated with respect to x

$$\mathcal{Y}(y,m_2,m_3) = \mathcal{R}_x(\tilde{F}_1,\tilde{F}_2) \quad (10)$$

is[1] the polynomial of the degree 34 in y . For any specialization of parameters m_2 and m_3, it is possible to find the exact number of real zeros and to localize the latter in the ideology of symbolic computations, e.g., via the Sturm series construction or via Hermite's method [5]. For instance, there exist 2 stationary points

$$\mathfrak{S}_1 \approx (2.666216, 1.234430), \quad \mathfrak{S}_2 \approx (2.744834, 3.244859)$$

for the case $m_2 = 2, m_3 = 2$, and 4 stationary points

$$\mathfrak{S}_1 \approx (1.941246, 2.552370), \mathfrak{S}_2 \approx (2.655622, 1.638871), \mathfrak{S}_3 \approx (3.330794, 2.826444),$$

and

$$\mathfrak{N} \approx (2.552939, 2.271691)$$

for the case $m_2 = 2, m_3 = 4$. Hereinafter we denote by $\mathfrak{S}$ the saddle-type stationary point and by $\mathfrak{N}$ the minimum point.

In order to find the boundary in the parameter (m_2, m_3)-plane between the two distinct qualitative pictures — i.e. two vs. four stationary points — let us find the *discriminant curve*. Any pair of bifurcation values corresponds to the case when at least one stationary point becomes degenerate, i.e. if these bifurcation values are perturbed somehow, this stationary point either splits into (at least) two ordinary, nondegenerate stationary points or disappears at all. Therefore these bifurcation values for parameters can be found from the condition of changing the number of real solutions of the system (9). Hence, the bifurcation values correspond to the case when the multiple zero for the polynomial (10) appears. This condition is equivalent to vanishing of the discriminant

$$\mathcal{D}_y(\mathcal{Y}) = \mathcal{R}_y(\mathcal{Y}, \mathcal{Y}'_y) . \quad (11)$$

---

[1] On excluding an extraneous factor.

This is a huge polynomial, which can be factored over $\mathbb{Z}$ as

$$\Xi^2(m_2, m_3)\Psi(m_2, m_3) \quad \text{with} \quad \deg \Xi = 444, \deg \Psi = 48 .$$

The condition $\Xi(m_2, m_3) = 0$ corresponds to the case where the multiple zero for (11) appears due to the coincidence of the values of y-components for a pair of distinct solutions of the system (9) while the condition

$$\Psi(m_2, m_3) = 0 \tag{12}$$

corresponds in the (m_2, m_3)-plane to the case of appearance of at least one degenerate zero for (9). The polynomial $\Psi(m_2, m_3)$ is an even one with respect to the involved parameters, and its expansion in powers of these parameters contains 325 terms. The complete expression can be found in [12], while here we demonstrate just only its terms of the highest and the lowest orders:

$$\Psi(m_2, m_3) = 3^{36}(64\,m_3^2 + 192\,m_2 m_3 + 169\,m_2^2)^5(64\,m_3^2 - 192\,m_2 m_3 + 169\,m_2^2)^5 \times$$

$$(28561\,m_2^4 + 19968\,m_2^2 m_3^2 + 4096\,m_3^4)^7$$

$$+ \ldots$$

$$+ 2^2 \cdot 3^{31} \cdot 17^{40}(5545037166327\,m_2^4 - 161882110764644\,m_2^2 m_3^2 + 1656772227072\,m_3^4)$$

$$+ 2^3 \cdot 3^{36} \cdot 17^{44}(51827\,m_2^2 + 28112\,m_3^2) + 3^{36} \cdot 17^{48} .$$

Drawing out the 48th order algebraic curve (12) is looking like an impossible mission. Fortunately[2], we have succeeded to do this. Since we are dealing with positive values of parameters, in Fig. 1 we present only the 4 "arrowhead"-looking branches of the curve lying in the first quadrant of the (m_2, m_3)-plane. Only one of these branches is the true bifurcation curve — the one presented in Fig. 2. The coordinates of its "vertices", i.e. singular points, are as follows:

$$M_1 \approx (1.812918, 2.575996), M_2 \approx (2.886962, 5.667175),$$

$$M_3 \approx (1.236728, 3.556856) .$$

The values of the parameters lying in the interior of this branch, i.e. satisfying the inequality $\Psi(m_2, m_3) < 0$, correspond to the case of existence of (precisely) 4 stationary points for the potential, while those lying outside this branch — to that of 2 stationary points.

It is worth mentioning, that for the case of existence of 4 stationary points, one of them is necessarily the minimum point for the potential, or, in terms of the differential equation theory, it gives the (asymptotically) stable equilibrium position for the vector field generated by the given configuration of charges. The condition $\Psi(m_2, m_3) < 0$ can therefore be tackled as a *necessary* one for the existence of the minimum (stable) point for the potential function; it should

---

[2] Thanks to the open-source mathematical software system **Sage**, http://www.sagemath.org.

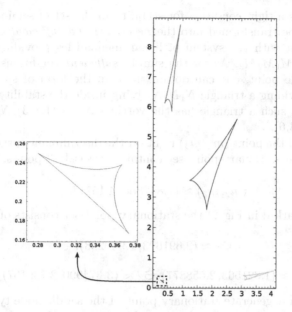

Fig. 1. Discriminant curve

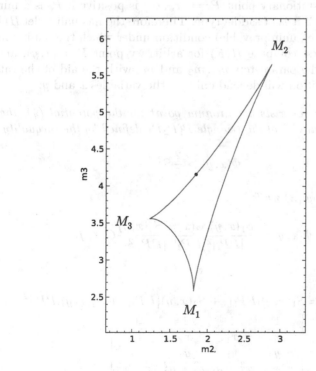

Fig. 2. Stability domain in parameter plane

be considered as a tight condition for stability in the set of semi-algebraic conditions. It can be transformed into the *necessary and sufficient* conditions via supplementing it with the system of linear inequalities providing the interior of the triangle $M_1 M_2 M_3$. As for the simple *sufficient* conditions for the existence of a stable point, one can obtain them in the form of system of linear inequalities providing a triangle $N_1 N_2 N_3$ lying inside the stability domain. For instance, one of such a triangle has the vertices $N_1 = (1.8, 3)$, $N_2 = (2.6, 5.1)$ and $N_3 = (1.4, 3.6)$.

The choice of the point (m_2, m_3) right on the discriminant curve corresponds to the case when stationary point set contains precisely 3 points. For instance, for

$$(m_2, m_3) \approx (1.842860, 4.157140)$$

(this point is marked in Fig. 2) the stationary point set consists of

$$\hat{\mathfrak{S}}_1 \approx (2.691693, 1.930238), \tag{13}$$

$$\mathfrak{S}_2 \approx (1.821563, 2.558877), \mathfrak{S}_3 \approx (3.374990, 2.739157),$$

with $\hat{\mathfrak{S}}_1$ being a degenerate stationary point of the saddle-node type.   □

The next challenging problem is that of localization of the stationary points for the potential (4). The determinant of the Hessian matrix $H(F)$ computed for this function at a stationary point $P_* = (x_*, y_*)$ is positive if P_* is a minimum point and negative if it is of saddle type. Therefore the inequality $\det H(F) > 0$ specifies a sharp (i.e., unimprovable) condition under which the minimum point exists. One may first compute $H(F)$ for arbitrary point $P = (x, y)$, and after that replace in it the parameters m_1, m_2 and m_3 with the aid of the ratio (7). The resulting condition will depend only on the variables x and y:

Theorem 2. *If there exists a minimum point for the potential (4), then it is located in the domain* $\mathbb{M}$ *of the triangle* $P_1 P_2 P_3$ *defined by the inequality*

$$\Phi(x, y) > \frac{2}{9} S^2 . \tag{14}$$

Here S is defined by (3) while

$$\Phi(x, y) = \frac{S_1(x, y) S_2(x, y) S_3(x, y)}{|PP_1|^2 |PP_2|^2 |PP_3|^2} C(x, y) \tag{15}$$

where

$$C(x, y) = S_1(x, y)|PP_1|^2 + S_2(x, y)|PP_2|^2 + S_3(x, y)|PP_3|^2$$

$$\equiv \begin{vmatrix} 1 & 1 & 1 & 1 \\ x & x_1 & x_2 & x_3 \\ y & y_1 & y_2 & y_3 \\ x^2 + y^2 & x_1^2 + y_1^2 & x_2^2 + y_2^2 & x_3^2 + y_3^2 \end{vmatrix}$$

and $\{S_j(x,y)\}_{j=1}^3$ are defined by (6). Conversely, any point $P_* = (x_*, y_*)$ lying in $\mathbb{M}$ is a minimum point for the potential (4) with any specialization of charges m_1, m_2, m_3 proportional to the values

$$m_1^* = S_1(x_*, y_*)|P_*P_1|^3, \quad m_2^* = S_2(x_*, y_*)|P_*P_2|^3, \quad m_3^* = S_3(x_*, y_*)|P_*P_3|^3.$$

For the proof of this theorem we refer to [11].

Example 2. Find the domain $\mathbb{M}$ of possible minimum point location for the configuration from Example 1.

Solution. Here[3] $S = 20$ and

$$\Phi(x,y) = \frac{16(28 - 5x - 3y)(5x - y - 4)(y - 1)(-52 + 30x + 32y - 5x^2 - 5y^2)}{((x-1)^2 + (y-1)^2)((x-5)^2 + (y-1)^2)((x-2)^2 + (y-6)^2)}.$$

The domain $\mathbb{M}$ is located inside the oval of the 6th order algebraic curve displayed in Fig. 3.

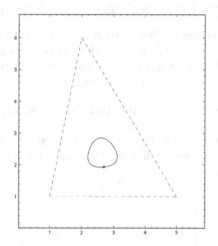

Fig. 3. Domain of possible minimum point location

One might expect that the point chosen on the curve corresponds to such values of parameters m_2 and m_3 that provide the degeneracy property of a stationary point for the potential. This is indeed the case: the one-to-one correspondence can be established between the points on this curve and those on the curve (12). For instance, the point marked on the curve in Fig. 3 is a degenerate stationary one for the potential (4) with $m_1 = 1, m_2 \approx 1.842860, m_3 \approx 4.157140$; its coordinates (13) have appeared in solution of Example 1.   □

To conclude the treatment of the three point case, let us consider the configuration of equal charges with one of their placement variable .

---

[3] In the article [11], the expressions for S and Φ are provided with typos.

Example 3. Let $P_1 = (0,0), P_2 = (1,0), P_3 = (x_3, y_3)$ and $m_1 = m_2 = m_3 = 1$. Analyse the structure of the set of stationary points of the function (4).

Solution. The idea is similar to the solution of Example 1. We skip the intermediate computations and present the final result: the discriminant curve in (x_3, y_3)-parameter plane is given implicitly as

$$\Theta(x_3, y_3) = 0 . \tag{16}$$

Here $\Theta(x_3, y_3)$ is the 76th order polynomial with respect to both coordinates. Its complete expression can be found in [12], while here we demonstrate only the terms of the highest and the lowest orders:

$$\Theta(x_3, y_3) =$$

$$2^{36} {\cdot} 3^{42} (9\,x_3^2 + 8\,y_3^2)^6 (x_3^2 + y_3^2)^{32} - 2^{37} {\cdot} 3^{42}\, x_3 (155\,y_3^2 + 171\,x_3^2)(9\,x_3^2 + 8\,y_3^2)^5 (x_3^2 + y_3^2)^{31}$$

$$+\ldots$$

$$-2^{37} {\cdot} 3^{42}\, x_3 (155\,y_3^2 + 171\,x_3^2)(9\,x_3^2 + 8\,y_3^2)^5 (x_3^2 + y_3^2) + 2^{36} {\cdot} 3^{42} (9\,x_3^2 + 8\,y_3^2)^6 (x_3^2 + y_3^2) .$$

(There are no terms of degree lesser than 14.) The curve (16) is symmetric with respect to the line $x_3 = 1/2$ and consists of two branches also symmetric with respect to the x_3-axis; one of these branches is displayed in Fig. 4. The coordinates of its singular points are as follows[4]:

$$Q_1 \approx (0.398295, 0.798718), \quad Q_2 \approx (0.601705, 0.798718), Q_3 \approx (0.5, 1.002671) .$$

For the point P_3 placed inside this curve ($\Theta(x_3, y_3) < 0$), the potential (4) possesses 4 stationary points, while for the point P_3 lying outside — just 2

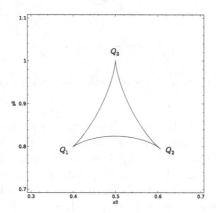

Fig. 4. The "upper" branch of the curve (16).

---

[4] Thus, one should not be misled by the visual illusion: the triangle $Q_1 Q_2 Q_3$ is not an equilateral one!

stationary points. Let us illuminate this statement considering the specialization $x_3 = 1/2$ which corresponds to the case of the isosceles triangle $P_1P_2P_3$. The corresponding stationary point set is symmetric with respect to the triangle median line $x_3 = 1/2$. One has

$$\Theta(1/2, y_3) \equiv \frac{1}{2^{48}} (65536\, y_3^8 + 16384\, y_3^6 - 13824\, y_3^4 - 15552\, y_3^2 - 2187)$$
$$\times\, (16384\, y_3^8 + 4096\, y_3^6 - 6912\, y_3^4 - 11664\, y_3^2 - 2187)^3 T^2(y_3)\ .$$

Here $T(y_3)$ denotes an even polynomial of the degree 22 without real roots, i.e. $T(y_3) \neq 0$ for $y_3 \in \mathbb{R}$. The roots of the remained factors are the bifurcation values for y_3, and we restrict ourselves here only by positive values:

$$y_3^* \approx 0.824539 \text{ and } y_3^{**} \approx 1.002671.$$

For the choice $y_3 \in (y_3^*; y_3^{**})$ the stationary point set consists of 4 points. For instance, if $y_3 = 1$ then these points are

$$\mathfrak{S}_1 \approx (0.520962,\ 0.424850),\, \mathfrak{S}_2 \approx (0.479037,\ 0.424850),\, \mathfrak{S}_3 \approx (0.5,\ 0.075682),$$

$$\mathfrak{N} \approx (0.5,\ 0.423647).$$

For $y_3 \in (0; y_3^*) \cup (y_3^{**}; \infty)$ the stationary point set consists of 2 points. For instance, if $y_3 = 1/2$ then these points are:

$$\mathfrak{S}_1 \approx (0.267236,\ 0.219775),\ \ \mathfrak{S}_2 \approx (0.732763,\ 0.219775)\ ;$$

while for the choice $y_3 = 3/2$ both stationary points lie on the line $x_3 = 1/2$:

$$\mathfrak{S}_1 \approx (0.5,\ 0.029031),\ \ \mathfrak{S}_2 \approx (0.5,\ 0.783949). \qquad\qquad \Box$$

3   Four Points

We now turn to the case of potential generated by configuration of 4 noncollinear charges $\{m_j\}_{j=1}^4$ placed at the points $\{P_j\}_{j=1}^4$

$$F(P) = \frac{m_1}{|PP_1|} + \frac{m_2}{|PP_2|} + \frac{m_3}{|PP_3|} + \frac{m_4}{|PP_4|}\ . \tag{17}$$

The idea of the proof of Theorem 1 can easily be extended to this case. System

$$\sum_{j=1}^4 \frac{m_j(x - x_j)}{|PP_j|^3} = 0,\ \ \sum_{j=1}^4 \frac{m_j(y - y_j)}{|PP_j|^3} = 0 \tag{18}$$

can be resolved — as a linear one — with respect to m_1 and m_2:

$$\begin{cases} m_1 S_3/|PP_1|^3 = m_3 S_1/|PP_3|^3 + m_4 S_4/|PP_4|^3, \\ m_2 S_3/|PP_2|^3 = m_3 S_2/|PP_3|^3 + m_4 S_5/|PP_4|^3. \end{cases} \tag{19}$$

Here S_1, S_2, and S_3 are defined by (6) while

$$S_4(x,y) = \begin{vmatrix} 1 & 1 & 1 \\ x & x_2 & x_4 \\ y & y_2 & y_4 \end{vmatrix}, \quad S_5(x,y) = \begin{vmatrix} 1 & 1 & 1 \\ x_1 & x & x_4 \\ y_1 & y & y_4 \end{vmatrix}.$$

Next, the squaring procedure can be applied to both equations of the system (19). In two steps this results in the algebraic equations of degree 28 in x and y:

$$\tilde{F}_1(x, y, m_1, m_2, m_3, m_4) = 0, \quad \tilde{F}_2(x, y, m_1, m_2, m_3, m_4) = 0 . \tag{20}$$

This result should be treated as an essential simplification in comparison with the squaring algorithm applied directly to system (18): the latter generates algebraic equations of the degree 72.

Example 4. Let $P_1 = (1,1), P_2 = (5,1), P_3 = (2,6), P_4 = (4,5)$. Analyse the structure of the set of stationary points of the function (17) for $m_1 = 1, m_2 = 2, m_3 = 4$ and for m_4 treated as parameter.

Solution. The given configuration of charges can be tackled as a perturbation of the 3 charges configuration from the solution of Example 1 with an extra charge placed at the position P_4.

We eliminate x variable from the system (20) with the aid of the resultant computation:

$$\mathcal{Y}(y, m_4) = \mathcal{R}_x(\tilde{F}_1, \tilde{F}_2) . \tag{21}$$

It can be factored over $\mathbb{Z}$ as

$$\mathcal{Y} \equiv W(m_4)G_1(y, m_4)G_2(y, m_4)(y-1)^{56}(y-5)^{16}(y-6)^{16}(4y^2 - 44y + 125)^{36} ;$$

here

$$W(m_4) \equiv m_4^{48}(m_4 - 5)^5(m_4 + 5)^5$$

and, generically, $\deg_y G_1 = 180, \deg_y G_2 = 156$.

The y-components of the zeros of the system (20) are among the zeros of $G_2(y, m_4)$. Although we have succeed to compute this polynomial, we have failed to find its discriminant with respect to the variable y. Therefore we are not able to provide one with the bifurcation values set for the parameter m_4. We have established that one of these values lies within the interval $(4/15; 3/10)$, and when the parameter m_4 passes through this value while decreasing, the number of stationary point increases from 3 to 5. For instance, one obtains

$$\mathfrak{S}_1 \approx (1.952957, 2.176070), \; \mathfrak{S}_2 \approx (4.239198, \; 2.677284), \; \mathfrak{S}_3 \approx (3.154287, \; 5.396890)$$

for $m_4 = 2$ and

$$\mathfrak{S}_1 \approx (1.988731, 2.474302), \; \mathfrak{S}_2 \approx (2.603988, 1.852183), \; \mathfrak{S}_3 \approx (3.593059, 2.883524),$$

$$\mathfrak{S}_4 \approx (3.566307, 5.178565), \mathfrak{N} \approx (2.560190, 2.031979)$$

for $m_4 = 4/15$.   □

In order to confirm Maxwell's conjecture for the case of $K = 4$ charges, we have generated about thirty variants of their configuration. The number of stationary points never exceeds 7.

Example 5. Find the stationary point set for the system of charges $m_1 = 1, m_2 = 3, m_3 = 1, m_4 = 3$ placed at $P_1 = (0,0), P_2 = (1/2,-1), P_3 = (1,0), P_4 = (1/2, 1)$ respectively.

Solution. The quadrilateral $P_1P_2P_3P_4$ is a rhombus, therefore the considered configuration of charges possesses two axes of symmetry, namely the lines $x = 1/2$ and $y = 0$. The symmetry property is inherited by the set of stationary points of the generated potential:

$$\mathfrak{N}_{1,2} \approx (0.5, \pm 0.194213),$$

$$\mathfrak{S}_1 = (0.5, 0), \mathfrak{S}_{2,3} \approx (0.316723, \pm 0.323720), \mathfrak{S}_{4,5} \approx (0.683276, \pm 0.323720).$$

Hence, at present we are unable neither to disprove Maxwell's estimation nor to ascertain its attainability.

4 Conclusions

Analytical approach for the investigation of the set of stationary points for the Coulomb potential function $F(P) = \sum_{j=1}^{K} m_j / |PP_j|$ in $\mathbb{R}^2$ was developed. The efficiency of the approach was illuminated for system of $K = 3$ and $K = 4$ charges in case when the values of charges as well as their coordinates are *specialized*, i.e. for the case when numerical values for these parameters are assigned, one can establish the exact number of stationary points and localize them within the given tolerance in a finite number of elementary algebraic operations. Moreover, for the case of $K = 3$ charges, it is possible to find the bifurcation picture in the domain of parameter *variation*. In all the examples we have treated Maxwell's conjecture was confirmed.

The case of $K \geq 4$ points in the space remains for further investigation. On extrapolating the length of outputs in examples treated in the paper, one may predict the growth of complexity in computation and analysis of this problem. On the other hand, it should be mentioned that all the computations for the examples have been performed on a standard configuration personal computer. Thus, the planned usage of specialized software implemented on a high-performance computer looks promising.

The proposed approach for construction of bifurcation diagrams can be applied for establishing the stability or ultimate boundedness conditions in the parameter space for wide classes of dynamical systems, such as treated in [1].

Acknowledgments. The authors are grateful to the anonymous referees for constructive suggestions and to Ivan Baravy for his help in drawing the figures. This work was supported by the St. Petersburg State University research grant # **9.38.674.2013**.

References

1. Aleksandrov, A.Y., Platonov, A.V.: On stability and dissipativity of some classes of complex systems. Autom. Remote Contr. **70**(8), 1265–1280 (2009)
2. Cox, D.A., Little, J., O'Shea, D.: Ideals, Varieties, and Algorithms. Springer, New York (2007)
3. Exner, P.: An isoperimetric problem for point interactions. J. Phys. A Math. Gen. **38**, 4795–4802 (2005)
4. Gabrielov, A., Novikov, D., Shapiro, B.: Mystery of point charges. Proc. London Math. Soc. **3**(95), 443–472 (2007)
5. Kalinina, E.A., Uteshev, A.Y.: Determination of the number of roots of a polynomial lying in a given algebraic domain. Linear Algebra Appl. **185**, 61–81 (1993)
6. Killian, K.: A remark on Maxwell's conjecture for planar charges. Complex Var. Elliptic Equ. **54**, 1073–1078 (2009)
7. Maxwell, J.C.: A Treatise on Electricity and Magnetism, vol. 1. Dower, New York (1954)
8. Peretz, R.: Application of the argument principle to Maxwell's conjecture for three point charges. Complex Var. Elliptic Equ. **58**(5), 715–725 (2013)
9. Tamm, I.: Fundamentals of the Theory of Electricity. Mir Publishers, Moscow (1979)
10. Uspensky, J.V.: Theory of Equations, pp. 251–255. McGraw-Hill, New York (1948)
11. Uteshev, A.Y., Yashina, M.V.: Stationary points for the family of fermat–torricelli–coulomb-like potential functions. In: Gerdt, V.P., Koepf, W., Mayr, E.W., Vorozhtsov, E.V. (eds.) CASC 2013. LNCS, vol. 8136, pp. 412–426. Springer, Heidelberg (2013)
12. Uteshev, A.Yu.: Notebook. http://pmpu.ru/vf4/matricese/optimize/coulomb_e

Application Control and Horizontal Scaling in Modern Cloud Middleware

Oleg Iakushkin[(✉)], Olga Sedova, and Grishkin Valery

Saint Petersburg State University, 7-9, Universitetskaya nab.,
St. Petersburg 199034, Russia
o.yakushkin@spbu.ru

Abstract. This work is focused on a number of standard communication patterns of distributed system nodes via messages. Certain characteristics of modern practically applied communication systems are considered. The conclusions are based on the practical development of collective communication strategy processing services and the theoretical basis drawn in the course of testing a number of distributed system prototypes. Development trends of service oriented architecture in the field of interservice communications are considered, including the development tendencies of AMQP and ZMTP protocols.

Problems arising during the design and development of such systems from the horizontal scaling standpoint are specified. The problem of long term control is highlighted in the course of considering issues of data consistency between nodes, availability and partition tolerance. The process of changing workload distribution in a horizontally scaled system is described and issues of fault tolerance of the system in general and its nodes in particular are raised. A way of workload scaling by means of defining an evaluation criterion of node load determined by the system's business logic and not by the characteristics of the communications level is offered. The efficiency of this approach is shown, with long term control systems used as an example.

Keywords: Cloud middleware · Communication patterns · Horizontal scaling · Zeromq

1 Introduction

This article is devoted to the issues of application control and solution of horizontal scaling tasks in Cloud Middleware. The main difficulty lies in the impossibility to perform any dynamical balancing in a cloud. Yet, this reason falls into quite a range of particular issues.

Fundamentally, the problem is that standard profilers cannot operate under TCP/IP network connections. In other words, the traditional architectures of complex applications cannot be directly transferred into a cloud. Standard working practices are unrealizable, so there is a need for new solutions.

M.L. Gavrilova and C.J. Kenneth Tan (Eds.): Trans. on Comput. Sci. XXVII, LNCS 9570, pp. 81–96, 2016.
DOI: 10.1007/978-3-662-50412-3_6

By following the logic of the specified task research, we have analyzed a number of solutions suggested by our colleagues that are applicable for large applications' work in cloud environments [1–3]. We have studied their advantages and disadvantages from the horizontal scaling perspective. The analysis has shown their inefficiency when it comes to solving the task of continuous delegation of control between the communicating components in the form of a dialogue. That is why we have suggested a new unconventional solution to control such applications.

Let us note that various tasks are solved in clouds. Three of them are evident upon initial consideration: a great number of small tasks such as mass calculations, processing of large arrays of data and calculation of one big task an integral control system changing its state upon the introduction of each new controlling action. The requirements for balancing and scaling of the systems solving these tasks are different. An additional circumstance is that, if we disregard the first case which is well-scalable even under the Guide, it is impossible to work in cloud structures without any means of communication.

We use different tools depending on which task we have to solve. Let us jump ahead and briefly describe the solution we suggest. Within the framework of a cloud system, the issues of dynamical workload balancing between the nodes become quite burning ones if it concerns tasks connected with intensive node communication, when the controlling elements are distributed between the virtual machines and the processes are completely virtualized. This problem is essential both for the Big Data processing tasks and for solving large system modeling tasks. We suggest that elements balancing the workload within the system framework at the communication level should be introduced into the inter-node communication logic, thus enabling to control the node load depending on the business logic requirements, to choose expansion strategies of the consumed resources and to implement new node communication patterns.

In particular, we suggest creating a common module for communication flow control. This module should allow connecting components of the business logic level that can work with workload balancing schemes. That enables us to use, when working with a particular infrastructure, the available system state monitoring means such as Zabbix or Globula by having optimally redistributed the information flows onto free nodes. However, all the inter-node communications remain direct, thus enabling the system architect to implement an optimal configuration of the distributed system communications. The architecture, including centralized orchestration of control flows in the system and direct communication of the nodes, combines the flexibility of architectural solutions of P2P networks and the convenience of broker message transmission systems.

Our work considers the issue of the suggested system operation from the perspective of message transmission patterns and describes the testing and approval thereof with evidence from the Computing Resource Center of St. Petersburg State University and with partial support of SPbU (Saint Petersburg State University) grant 0.37.155.2014 and Russian Foundation for Basic Research (project no. 16-07-01113).

1.1 Related Work

Horizontal scaling requires a thorough elaboration of communication between distributed system nodes. The messages about actual changes in the system and about nodes communication are transmitted between the components at the communication level and this process is described by a number of message transmission patterns. The pattern ideology description includes workload distribution schemes used in horizontal scaling of the system. Analysis of various communication schemes between the distributed environment nodes is a crucial development trend for the field of service oriented architecture (SOA).

Upon the development of particular compute nodes (when powerful GPGPUs and other coprocessors appeared), calculation distribution for computation-intensive tasks between the system nodes has been growing in difficulty in terms of workload balancing. The access from the coprocessor to CPU via the PCI bus will always be quicker than an inter-node connection using InfiniBand and Ethernet. At the same time, today coprocessors are closing the gap between them and CPUs at a great speed in terms of the quantity of available instructions, so that certain tasks equivalent to those for the CPU can be solved using them. This technology development thrust is characterized by the fact that more and more parts of distributed algorithms' work are delegated to particular system units while inter-node communications specialize in business logic operation to a greater extent. The support of applications consuming more and more resources has stimulated the development of middleware enabling user applications to detach themselves from the resources they are running upon [4, 5]. In this regard, cloud middleware is being developed at major scientific centers all over the world. Today, such middleware is an integral part of any GRID and cloud infrastructure [6, 7].

The increasing performance of compute resources influences the development thrust of high-performance systems: software as a service (SaaS) is actively developing and, together with it, service oriented architecture (SOA) [8]. SOA is based upon loosely-coupled components, that meaning a growing number of applications have to solve tasks of a number of users simultaneously by providing different access levels to their resources [9]. The increase of service logic and extraction of system components into different applications not directly linked to each other raises the issue of communication and control. Cloud system infrastructure consists of a cloud network (element-connecting) and cloud middleware being responsible for their control [7]. The most popular solution here is exercising control over distributed system components via transmission of messages, thus enabling service components to be used in various execution contexts [10]. There are many open-source (RabbitMQ, ActiveMQ, Qpid, etc.) [11–14] and proprietory (Windows Azure Service Bus, Amazon SQS, etc.) [15, 16] implementations of the communication layer in cloud infrastructures. In Fig. 1(a) a client-brocer interection is presented. Broker implementations using the AMQP protocol are the most popular ones. Figure 1(b) shows how non-broker P2P communication implementations, that connect components directly, have been actively developing over the most recent five years [17]. ZeroMQ is the best example of such a P2P implementation [18]. Analysis of such systems' efficiency is an essential developing field: various issues of network latency, system nodes' availability and data consistency are being studied [19–21]. The most prominent study comparing the existing implementations of communication layers is probably the work of CERN scientists presented at the 13th International Conference on Accelerator and Large Experimental Physics Control Systems in 2011 [22]. It has revealed a technological superiority of the ZeroMQ P2P system.

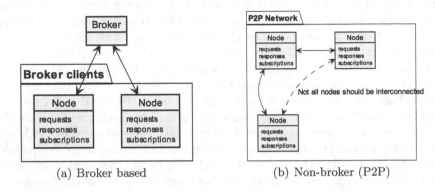

(a) Broker based   (b) Non-broker (P2P)

Fig. 1. Nodes communication

CERN has recently hosted a conference devoted to remote device access (RDA) middleware based on a P2P network formed on the ZeroMQ transport (RDA3) [23]. Among other things, this conference featured a discussion concerning a transition from the old CORBA-based control system that has almost worked itself out to the up-to-date RDA system which is ZeroMQ-based [12,24]. Issues concerning the current CERN RDA architecture and its capability to fit into ZeroMQ were studied [25]. It should be emphasized that in 2012 the International Conference on Computing in High Energy and Nuclear Physics featured a presentation of RDA2, which is also ZeroMQ-based and yet works with CORBA on equal terms, not excluding it [26].

Today, broker MQ systems are an integral part of the AMQP protocol that appeared in 2006 [27] as a result of JPMorgan Chase group work that included Cisco Systems, IONA Technologies, iMatix, Red Hat and Transaction Workflow Innovation Standards Team. That gave an impetus to broker systems and thus Apache Qpid, RabbitMQ, StormMQ were founded. The AMQP protocol was actively developing, its latest specification being presented in 2011 [28].

The following modern design patterns of service P2P systems connected via messages were among the first theoretical descriptions: Request-Response, Subscribe-Push, Probe and Match (now known as Survey-Respondent). These descriptions were published in 2007 under the authorship of Duane Nickull, Laurel Reitman and James Ward: Service Oriented Architecture (SOA) and Specialized Messaging Patterns [29].

The most detailed description of distributed system design patterns is given in the book by Pieter Hintjens, ZeroMQ Messaging for Many Applications, published in 2013. He is also the author of the ZeroMQ P2P communication library [30]. Yet, many of the patterns he specifies require certain additional description in terms of the horizontal scaling of a system being designed. This work is devoted to this specific task.

2   Service Communication Patterns

The fundamental limitations of distributed systems described by Eric Brewer in the CAP theorem [31] demonstrate that it is impossible to create a distributed system that could simultaneously provide all of the following guarantees: data consistency in all the nodes of the system, continuous availability of services and partition tolerance. The growing popularity of the BASE approach to system design was the main response to that [32]. It reduced the main requirements of the CAP theorem to provision of basic availability, unstable state between the nodes and eventual consistency.

A practical implementation of a distributed service-providing BASE system requires a number of things from the communication layer between the components: getting a message delivery status report taking into account the time frame of message life cycle; responding to the sender's message; broadcasting a message. With such transport properties, the system architect can evaluate any time delays appearing in case of components failure.

Let us consider the existing message transmission patterns in terms of system control opportunities they provide.

2.1   Publish-Subscribe

Messages are created by the publisher and sent to all of the subscribers. Such a system enables the publisher to inform the subscribers of any new data. This approach to data transmission is also called the events system. The main characteristic of this system is support of a number of subscribers that is not predefined and can change in the course of the system's operation. A practical application of the Publish-Subscribe pattern is clearly illustrated in the work devoted to the monitoring of cluster system [33] as well as in the article about quality assessment of service systems' operation [34].

In terms of horizontal scaling this leads us to the fact that, upon creating the publisher's clone, we should subscribe to its messages only those components which have not previously been subscribed to the original publisher. That means that without increasing the number of subscribers in the system it makes no sense to increase the number of publishers. To synchronize the publishers that have been scaled, the synchronization of the initial state is sufficient if there are no external dependencies on other components of the system (Fig. 2). This messaging pattern is supported by the AMQP and ZMTP protocols and its implementations can be found in almost all modern communication systems.

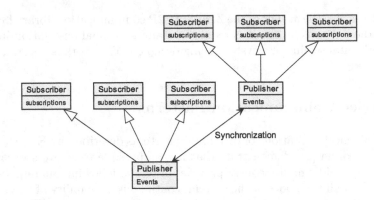

Fig. 2. Scaled Publish-Subscribe

2.2   Request-Reply

This is an operations system where the sender receives a reply upon their request. It is used to create systems of remote procedure call and is known as the client/server pair. The service system providing remote access to GPU resources is a good example of a practical application of the Request-Reply pattern [35]. The number of clients in it may increase depending on the number of servers since the operations are limited in time and yield an outcome that simultaneously changes the state of all components involved. Such operations require at least twice as much time because the messages participating in an operation are bidirectional. In Fig. 3 is shown a system that can be easily scaled by increasing the number of nodes; yet, when clients' operations are capable of changing the state of the server, additional synchronization will be needed between the server, whose state has been changed, and its replicas (). This is quite time-consuming and may lead to asynchronous behavior of the state of the system nodes and eventually to unavailability of the system as shown in Fig. 4. This messaging pattern is implemented in all broker and P2P message transmission systems.

2.3   Pipeline

This is a scheme of component connection construction based on progressive processing and transmission of data. In terms of architecture the solution can be represented as a connected directed acyclic graph. In other words, control is exercised by transmitting messages between the handlers in one direction, thus forming the application logic. Such an approach, dividing programs into components distributed on different nodes, allows us to create complex graphs of task execution with horizontal scaling localized on certain parts of program execution. This method has become widely popular in visual modeling environments of complex systems. Creation of compatible components requires unification of communication interfaces from software engineers. P2P communication systems are used for operation of such distributed systems. In Fig. 5 we show how the

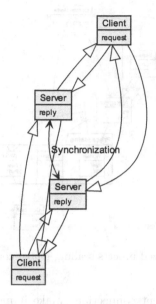

Fig. 3. Scaled Request-Reply

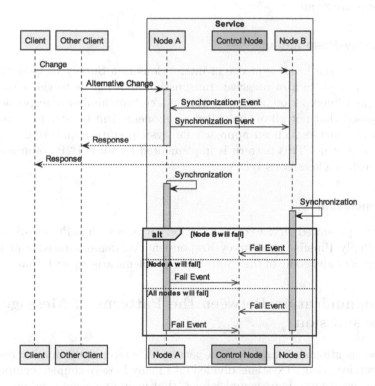

Fig. 4. Nodes fail due to synchronization issues

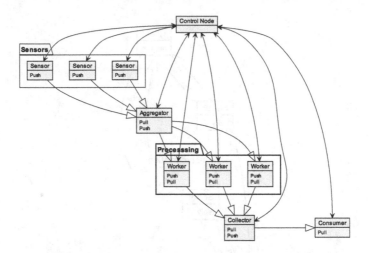

Fig. 5. Localized process scaling, global controll node

complex graphs of component connections make it more difficult to create an overall picture of system operation - certain supporting monitoring and control components are required.

2.4 Survey-Respondent

Figure 6 shows the basic sequence of interactions in a Survey-Respondent pattern. This is a pattern of message transmission allowing us to get a summary during a predefined period of time by polling a certain number of respondents (). It is supposed that not all of the polled components will be able to respond. It is convenient to use such an approach for system state monitoring and node statistics collecting. This pattern is implemented by some P2P communication systems such as Crossroads I/O.

2.5 Conclusion

In Sect. 2 we presented service communication patterns such as Publish-Subscribe, Request-Reply, Pipeline and Survey-Respondent. We demonstrated implementation differences and some of the control flow problems arising with that.

3 Communication Between the Patterns of Message Transmission

The above-mentioned communication patterns are often used simultaneously in modern service oriented systems divided into many loosely-coupled components. There they evolve into more complex ones, that including events and operations, increasing the number of closely connected elements and possible deadlocks.

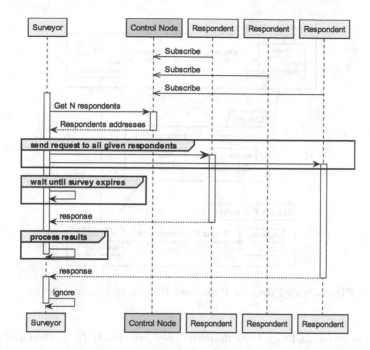

Fig. 6. Survey-Respondent sequence diagram

This leads to new requirements to the communication layer for optimal horizontal scaling: balancing of workload distribution and supporting at the channel level of message transmission concerning the existence of subsystems responsible for control graph consistency.

3.1 Events and Operations

In Fig. 7 shows combination of Request-Reply and Publish-Subscribe patterns. This are the most widespread examples of such combination. This combination does not specify any particular requirements to system horizontal scaling except for those imposed by communication patterns. Yet, it raises a burning issue of mutual node deadlocks shown in Fig. 8: in the course of waiting for the events and brings various traffic to the communication level. The latter, in its turn, greatly increases the estimation complexity of bandwidth capacity of connections between nodes.

3.2 Operation Chains

Partial delegation of control from one component to another is widely used in service systems. Such delegation is based on operation chains and implies an opportunity of the components to consciously finish their dialogue. The state of participating components can change on each step of this dialogue. That means that

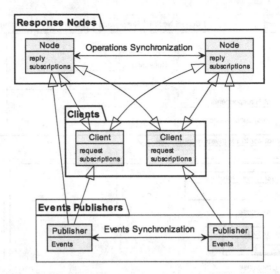

Fig. 7. Scaled Request-Reply and Publish-Subscribe mixture

until the components finish their dialogue they are closely connected and exist in an undefined state. Upon finishing the dialogue, the components can synchronize their states with the replicas, like it is with the operations. Unlike operations, such chains, changing the components' state, result in the fact that responses to parallel requests at the replicated nodes will be different if compared to the moment of entering the dialogue. Due to continuous and time-varying workload on the nodes in the course of the dialogue as well as due to the need of supporting communication between particular nodes, the scaling problem is a most pressing one. This problem has not yet been solved in modern communication systems based on AMQP-like protocols and in Peer-2-Peer solutions such as ZeroMQ, Crossroads I/O and NanoMsg.

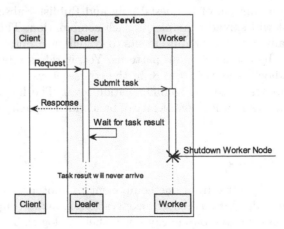

Fig. 8. Dealer is deadlocked waiting for event

3.3 Conclusion

In Sect. 3 we have shown some of the problems arising during communication between the patterns of message transmission such as operation chains and events and operations.

4 Our Contributions to the Area

We suggest a possible solution for the problem of horizontal scaling of a system with control dialogues. Our solution is based on introducing the node load criterion. This approach enables us to solve the task of resource distribution during horizontal scaling of the system only slightly modifying the node communication architecture.

Let us describe modifications required to implement this solution. Let us mark the messages and system nodes connected to them during the dialogues. This will enable us to introduce a new evaluation criterion of node load reflecting the number of dialogues currently carried out by this node.

Then, unlike the traditional Round-Robin and Random Access strategies of workload distribution, the dispatching element of the communication layer will be able to properly observe which of the system elements is loaded with which dialogues. Analysis of this data, in its turn, will allow us to choose an optimal candidate for new message processing. As a result of using this criterion, the workload will be proportionally distributed between the nodes. Ideally this will entail optimal distribution of messages per the system elements in case of operation chains.

The suggested approach of delegating the system elements' load control to the business logic level. In Fig. 9 is shown sequence of commands required for an algorithm implementation. It enables us, by loading the communication layer with additional information, to significantly facilitate the overall workload balancing of the system.

On the Fig. 10 we compare message-processing efficiency of a heterogeneous system composed from three nodes: two equivalent and one with half of computational power. We assume that after overcoming highest possible message consumption rate node fails. We show how much messages are processed in the system throughout four iterations for balanced and Round-Robin scheduling. Each iteration increases load by thirty messages. Balanced approach will be able to suspend or redirect incoming messages when node consuming capacity limit is reached, while Round-Robin nodes will overflow.

The capabilities of the existing communication layers are still very limited today to support the dynamical control of node load distribution. This encourages large-scale research in this field.

The storage of additional information about the messages will be by all means complicating the communication protocol. Yet, some protocols, such as AMPQ and XMPP, support an unlimited number of message tags and show good performance on the existing implementations.

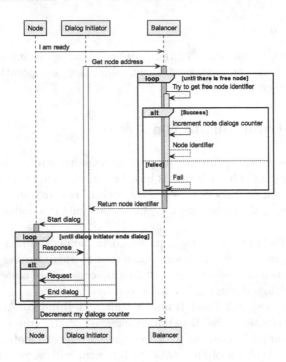

Fig. 9. Dialog sequence diagram

A negative factor introduced to the communication level by the additional information, in case of its incorrect usage, will put critical pressure upon the system; yet, it can be efficiently adjusted for each particular system to support its current needs.

4.1   Testing and Approval

The suggested solution was tested using the equipment provided by the resource center "SPbU Computing Facility" in the course of solving the task of modeling a dynamical route of agent groups in complex geographical conditions [36,37]. An ZeroMQ-based communication layer was created by us using the C# language. Apart from the main communication patterns, the developed layer supports operation chains as well as dynamical control of workload balancing. That allowed us to exercise an effective horizontal distribution of workload between the nodes responsible for the business logic of the agents' behavior and the nodes calculating their route and the dynamic world.

The delegation of agent control from the business logic system to the world calculation system for a step-by-step movement from the departure point to the destination became the operation chain in this system. In the course of this movement the business logic system could decide to change the destination. Such an approach enables us to delegate control for a prolonged period of time,

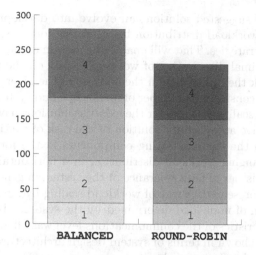

Fig. 10. Processed messages (in thousands / second)

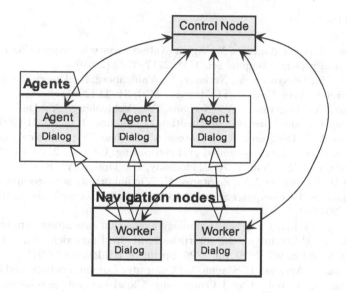

Fig. 11. Dialog connections distribuition inside running system

preserving at the same time an informative dialogue and providing optimal distribution of workload on the system components involved (Fig. 11).

5 Conclusion

For high-quality horizontal scaling of modern distributed systems, the collection and analysis of loading of communication systems' elements is a particularly essential task. The suggested solution can evolve into development of an interface for modeling workload distribution between the nodes using various workload distribution strategies. That will enable the system architect to define their own criteria of optimal distribution of workload based on the meta-data of the messages and check their efficiency in the context of the given system.

The article has considered a number of means of cloud system combining and scaling. Horizontal scaling schemes for the existing solutions have been suggested as well as an effective approach for solution of the task of continuous delegation of control between the communicating components in the form of a dialogue. The process of changing workload distribution in a horizontally scaled system was described and issues of fault tolerance of the system in general and its nodes in particular were presented. A way of workload scaling by means of defining an evaluation criterion of node load determined by the system's business logic and not by the characteristics of the communications level was offered. The efficiency of this approach is shown in terms of system design architecture, with long term control systems used as an example.

References

1. Degtyarev, A.B., Logvinenko, Y.: Agent system service for supporting river boats navigation. Procedia Comput. Sci. **1**(1), 2717–2722 (2010)
2. Bogdanov, A., Degtyarev, A., Nechaev, Y., Valdenberg, A.: Design of telemedicine system architecture. Healthc. IT Manage. **1**(2), 31–33 (2006)
3. Bogdanov, A., Degtyarev, A., Nechaev, Y., Valdenberg, A.: Design of high-performance telemedicine system. Healthc. IT Manage. **1**(1), 29–31 (2006)
4. Bogdanov, A.V., Degtyarev, A.B., Mareev, V., Nechaev, Y.: Flexible dynamic pooling of resources or service-oriented grid computing. Inf. Soc. **2**, 61–70 (2012)
5. Gankevich, I., Gaiduchok, V., Gushchanskiy, D., Tipikin, Y., Korkhov, V., Degtyarev, A.B., Bogdanov, A.V., Zolotarev, V.: Virtual private supercomputer: Design and evaluation. In: Computer Science and Information Technologies (CSIT), IEEE, pp. 1–6 (2013)
6. Goedicke, M., Zdun, U.: A key technology evaluation case study: Applying a new middleware architecture on the enterprise scale. In: Emmerich, W., Tai, S. (eds.) EDO 2000. LNCS, vol. 1999, pp. 8–26. Springer, Heidelberg (2001)
7. Doddavula, S., Agrawal, I., Saxena, V.: Computer Communications and Networks. In: Mahmood, Z. (ed.) Cloud Computing. Cloud computing solution patterns: Infrastructural solutions, pp. 197–219. Springer, London (2013)
8. Josuttis, N.: SOA in Practice. O'reilly, Sebastopol (2007)
9. Ghag, S.S., Bandopadhyaya, R.: Technical strategies and architectural patterns for migrating legacy systems to the cloud. In: Mahmood, Z., Saeed, S. (eds.) Software Engineering Frameworks for the Cloud Computing Paradigm. Computer Communications and Networks, pp. 235–254. Springer, Heidelberg (2013)
10. Huhns, M.N., Singh, M.P.: Service-oriented computing: Key concepts and principles. Int. Comput. IEEE **9**(1), 75–81 (2005)
11. Petcu, D., Rak, M.: Open-source cloudware support for the portability of applications using cloud infrastructure services. In: Mahmood, Z. (ed.) Cloud Computing. Computer Communications and Networks. Springer, Heidelberg (2013)

12. Yastrebov, I.: Rda3 high-level - api & architecture (2013). http://indico.cern.ch/getFile.py/access?contribId=3&resId=1&materialId=slides&confId=259755
13. Snyder, B., Bosnanac, D., Davies, R.: ActiveMQ in action. Manning (2011)
14. Videla, A., Williams, J.J.: RabbitMQ in action. Manning (2012)
15. Amazon, S.: Team, building scalable, reliable amazon ec2 applications with amazon sqs (2008). http://sqs-public-images.s3.amazonaws.com/Building_Scalabale_EC2_applications_with_SQS2.pdf
16. Microsoft: Windows azure service bus (2012). http://www.windowsazure.com/en-us/develop/net/fundamentals/hybrid-solutions/
17. Prinz, V., Fuchs, F., Ruppel, P., Gerdes, C., Southall, A.: Adaptive and fault-tolerant service composition in peer-to-peer systems. In: Meier, R., Terzis, S. (eds.) DAIS 2008. LNCS, vol. 5053, pp. 30–43. Springer, Heidelberg (2008)
18. Piël, N.: Zeromq an introduction. Retrieved **6**(30), 2011 (2010)
19. Oudenstad, J., Rouvoy, R., Eliassen, F., Gjørven, E.: Brokering planning metadata in a P2P environment. In: Meier, R., Terzis, S. (eds.) DAIS 2008. LNCS, vol. 5053, pp. 168–181. Springer, Heidelberg (2008)
20. Schmid, M., Kroeger, R.: Decentralised QoS-management in service oriented architectures. In: Meier, R., Terzis, S. (eds.) DAIS 2008. LNCS, vol. 5053, pp. 44–57. Springer, Heidelberg (2008)
21. Wu, Q., Gu, Y.: Performance analysis and optimization of linear workflows in heterogeneous network environments. In: Preve, N.P. (ed.) Grid Computing. Computer Communications and Networks, pp. 89–120. Springer, Heidelberg (2011)
22. Dworak, A., Sobczak, M., Ehm, F., Sliwinski, W., Charrue, P.: Middleware trends and market leaders 2011. Technical report (2011)
23. Review of the controls middleware transport architecture and its use of zeromq (2013). http://indico.cern.ch/conferenceDisplay.py?confId=259755
24. Sliwinski, W.: Controls middleware renovation - technical overview (2013). http://indico.cern.ch/getFile.py/access?contribId=2&resId=1&materialId=slides&confId=259755
25. Lauener, J.: Rda3 transport (2013). http://indico.cern.ch/getFile.py/access?contribId=3&resId=1&materialId=slides&confId=259755
26. Dworak, A., Ehm, F., Charrue, P., Sliwinski, W.: The new cern controls middleware. J. Phys.: Conf. Ser. **396**, 012017 (2012). IOP Publishing
27. Vinoski, S.: Advanced message queuing protocol. Int. Comput. IEEE **10**(6), 87–89 (2006)
28. Group, A.W: Amqp v1.0. (2011). http://www.amqp.org/sites/amqp.org/files/amqp.pdf
29. Reitman, L., Ward, J., Wilber, J.: Service oriented architecture (soa) and specialized messaging patterns. A technical White Paper published by Adobe Corporation USA (2007)
30. Hintjens, P.: ZeroMQ: Messaging for Many Applications. O'Reilly (2013)
31. Brewer, E.A.: Towards robust distributed systems. In: PODC, vol. 7 (2000)
32. Pritchett, D.: Base: An acid alternative. Queue **6**(3), 48–55 (2008)
33. Focht, E., Jeutter, A.: AggMon: Scalable hierarchical cluster monitoring. In: Resch, M.M., Wang, X., Bez, W., Focht, E., Kobayashi, H. (eds.) Sustained Simulation Performance 2012, pp. 51–64. Springer, Heidelberg (2013)
34. Ivanović, D., Carro, M., Hermenegildo, M.: Constraint-based runtime prediction of SLA violations in service orchestrations. In: Kappel, G., Maamar, Z., Motahari-Nezhad, H.R. (eds.) Service Oriented Computing. LNCS, vol. 7084, pp. 62–76. Springer, Heidelberg (2011)

35. E. Duran, R., Zhang, L., Hayhurst, T.: Enabling GPU acceleration with messaging middleware. In: Abd Manaf, A., Sahibuddin, S., Ahmad, R., Mohd Daud, S., El-Qawasmeh, E. (eds.) ICIEIS 2011, Part III. CCIS, vol. 253, pp. 410–423. Springer, Heidelberg (2011)
36. Rao, J.S.: Optimization. In: Rao, J.S. (ed.) History of Rotating Machinery Dynamics. HMMS, vol. 20, pp. 341–351. Springer, Heidelberg (2011)
37. Iakushkin, O.: Intellectual scaling in a distributed cloud application architecture: A message classification algorithm. In: 2015 International Conference Stability and Control Processes in Memory of V.I. Zubov (SCP), pp. 634–637, October 2015

A Built-in Self-repair Circuit for Restructuring Mesh-Connected Processor Arrays by Direct Spare Replacement

Itsuo Takanami[1](✉), Tadayoshi Horita[2], Masakazu Akiba[2], Mina Terauchi[2], and Tsuneo Kanno[2]

[1] Tokyo, Japan
iftakanami@comet.ocn.ne.jp
[2] Polytechnic University, 2-32-1, Ogawanishimachi, Kodaira-shi,
Tokyo 187-0035, Japan
horita@uitec.ac.jp
http://www.uitec.jeed.or.jp/english/index.html

Abstract. We present a digital circuit for restructuring a mesh-connected processor array with faulty processing elements which are directly replaced by spare processing elements located at two orthogonal sides of the array. First, the spare assignment problem is formalized as a matching problem in graph theory. Using the result, we present an algorithm for restructuring the array in a convenient form for finding a matching by a digital circuit. Second, the digital circuit which exactly realizes the algorithm is given. The circuit can be embedded in a target processor array to restructure very quickly the array with faulty processing elements without the aid of a host computer. This implies that the proposed system is effective in not only enhancing the run-time reliability of a processor array but also such an environment that the repair by hand is difficult or a processor array is embedded within a VLSI chip where faulty processor elements cannot be monitored externally through the boundary pins of the chip, and so on. Third, the data about the array reliability considering not only faults in processors but also in that digital circuit are given, and then the effectiveness of our scheme is shown.

Keywords: Fault-tolerance · Mesh array · Self-repair · Built-in circuit · Graph theory

1 Introduction

Recently, high-speed and high-quality technologies for processing many kinds of information have become essential. It is expected that higher-speed and higher-quality technologies will become more and more necessary in the future. For these needs, how to realize high-speed, massively parallel computers has been studied in the literature. A mesh-connected processor array ("PA" for short) is a kind of form of massively parallel computers. Mesh-connected PAs consisting

© Springer-Verlag Berlin Heidelberg 2016
M.L. Gavrilova and C.J. Kenneth Tan (Eds.): Trans. on Comput. Sci. XXVII, LNCS 9570, pp. 97–119, 2016.
DOI: 10.1007/978-3-662-50412-3_7

of processing elements ("PE" for short) have regular and modular structures which are very suitable for most signal and image processing algorithms. One vulnerable feature of such arrays is that if a single PE fails to perform its intended function correctly, due to some physical defects, the entire computation may become erroneous. On the other hand, as VLSI technology has developed, the realization of parallel computer systems using multi-chip module (MCM) or wafer scale integration (WSI) has been considered so as to enhance the speed of the computers, decrease energy consumption and sizes, and so on. In such a realization, entire or significant parts of PEs and connections among them are implemented on a board or wafer. Therefore, the yield and/or reliability of the system may become drastically low if there is no strategy for coping with defects and faults. To restore the correct computation capabilities of the array, it must be restructured appropriately so that the defective PEs are eliminated from the computation paths, and the working PEs maintain correct logical connectivities between themselves. Various strategies to restructure a faulty physical system into a fault-free target logical system are described in the literature, e.g., [1–10]. Some of these techniques employ very powerful restructuring schemes ("RS" for short) that can repair a faulty array with almost certainty, even in the presence of clusters of multiple faults. However, the key limitation of these techniques is that they are executed in software programs to run on an external host computer and they cannot be implemented efficiently within a PA chip by using special-purpose circuits. If a PA with faulty PEs can be restructured using a built-in circuit, the system down time of the PA is reduced. Furthermore, the PA will become more reliable when it is used in such an environment that the repair by hand is difficult or the PA is embedded within a VLSI chip where faulty PEs cannot be monitored externally through the boundary pins of the chip, and so on.

As far as we know, the first attempts to develop RSs were made by Negrini et al. [11] and Sami and Stefanelli [12]. The design approach of their repair control scheme begins with a heuristic algorithm that involves only some local reconnection operations. Mazumder et al. [13] developed the automatic RS using a built-in self-repair circuit for mesh-connected PAs with spares at two orthogonal sides of the arrays by which faulty PEs are directly replaced. The authors also developed the automatic RSs [14–19] for mesh-connected PAs using $1\frac{1}{2}$-track-switch scheme ("1.5-TS" scheme for short) proposed by Kung et al. [1] where the PAs have two rows and two columns of spares to cope with faults of low reliable PEs. However, the restructuring in these automatic RSs are done in heuristic ways in selecting routes to spares. Besides the approaches above, Lin et al. [20] proposed the fault-tolerant path router with built-in self-test/self-diagnosis and fault-isolation circuits for 2D-mesh based chip multiprocessor systems. Collet et al. [21] also proposed the chip self-organization and fault tolerance at the architectural level to improve dependable continuous operation of multi-core arrays in massively defective nanotechnologies, where the architectural self-organization results from the conjunction of self-diagnosis and self-disconnection mechanism, plus self-discovery of routers to maintain the communication in the array.

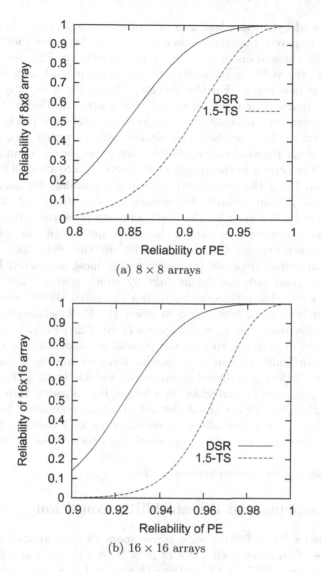

(a) 8×8 arrays

(b) 16×16 arrays

Fig. 1. Reliabilities of DSR and 1.5-TS mesh arrays with spares on one row and one column

On the other hand, if PEs are fairly reliable, it is expected that fewer spares will be enough for retaining reliability of an array so high and additional control schemes as well as control circuits will become simple. From this expectation, Takanami [22] proposed an automatic RS for 1.5-TS mesh arrays with spares on one row and one column. In this scheme, a faulty PE is locally replaced by it's neighboring nonfaulty PE so that physical distances between logical adjacent PEs are bounded by a constant. Besides the above, In [13,23,24], a faulty PE is directly replaced by a spare laid at a side of an array. In this scheme which is called a direct spare replacement scheme ("DSR" scheme for short) in the following, though physical distances between logical adjacent PEs are not bounded by a constant, the yields or reliabilities of arrays are much higher than those of 1.5-TS scheme (see Fig. 1). For this scheme, Mazumder et al. [13] developed a built-in self-repair circuit using an analog neural network. However, it not only restructures arrays in a heuristic way in selecting routes for replacing faulty PEs by spares, that is, the case that a repairable fault pattern (a set of faults) in DSR scheme is not repaired may occur, but also it has the disadvantages that (i) the biases to the neuron in the analog neural network are affected by the actual defect pattern, (ii) if the defect pattern is not repairable, the neural network will not be able to stop naturally by arriving a stable state, and (iii) the calculation of the transistor sizes for the exact amounts of currents through pull-up and pull-down transistor is critical in the neural net design where the synaptic influences between neurons are represented by electrical currents.

We present a digital circuit for restructuring mesh-connected PAs in DSR scheme. The circuit outputs signals that by which spares faulty PEs should be replaced while this replacement satisfies the repairability condition to be mentioned in Sect. 2. In Sect. 2, we mention the DSR scheme, formalize the strategies for deciding that by which spares faulty PEs should be replaced as a matching problem in graph theory, and present an algorithm for restructuring the arrays with faulty PEs in a convenient form for finding a matching by a digital circuit. In Sect. 3, a digital circuit which exactly realizes the algorithm is given. The circuit can be embedded in a target PA to restructure very quickly the array with faulty PEs without the aid of a host computer. In Sect. 4, the data from Monte Carlo simulations, considering not only faults in PEs but also in that digital circuit, are given, and then the effectiveness of our scheme is shown.

This paper is an extended version of [25].

2   DSR Scheme and Repairability Condition

Figure 2 shows a PA in DSR scheme where spare PEs are located on the upper and left sides of an array with a size of $N \times N$. A PE which is not a spare PE is called "non-spare PE". A PE at the i-th row and j-th column is denoted as p_{ij} $(0 \leq i \leq N, 0 \leq j \leq N, 1 \leq i + j)$ and hence the spare PEs on the upper and left sides are denoted as p_{0j}'s and p_{i0}'s, respectively. Then, if p_{ij} is faulty, it is directly replaced either by the spare PE p_{0j} or p_{i0}. Note that if a spare PE is faulty, it is considered to be replaced by itself.

Notation:

- If a faulty PE is replaced by a spare PE, it is said to be repaired, otherwise unrepaired.
- For a set of faulty PEs which is often called a fault pattern, if all the faults in the set can be repaired at the same time, the fault pattern is said to be repairable, otherwise unrepairable.

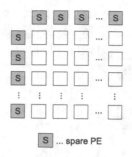

Fig. 2. PA in DSR scheme

For DSR scheme, we formalize a strategy for deciding that by which spares faulty PEs should be replaced as a matching problem in graph theory.

We construct the following bipartite graph G. $G = (V, E)$, where $V = V_f \cup (V_c \cup V_r)$, $V_f = \{p'_{ij} | 0 \le i \le N, 0 \le j \le N, 1 \le i + j, p_{ij}$ is faulty$\}$, $V_c = \{p_{10}, ..., p_{N0}\}$, $V_r = \{p_{01}, ..., p_{0N}\}$, $E = \{(p'_{ij}, p_{i0}) | 1 \le i \le N, 0 \le j \le N, p_{ij}$ is faulty$\} \cup \{(p'_{ij}, p_{0j}) | 0 \le i \le N, 1 \le j \le N, p_{ij}$ is faulty$\}$. G is called a compensation graph, V_f a (vertex) set of faulty PEs, $(V_c \cup V_r)$ a (vertex) set of spare PEs and E a set of edges implying replacement relation, respectively. Note that if a spare PE p is faulty, p' is in V_f and (p', p) is in E. Figure 3(a) shows an array in DSR scheme with faulty PEs expressed by "×"s, and (b) shows it's compensation graph which consists of three connected components CC_1, CC_2 and CC_3.

It is clear that the following holds.

Lemma 1. *A set of faulty PEs V_f is repairable in DSR scheme if and only if there exists a matching from V_f to $(V_c \cup V_r)$. For such a matching M, faulty p_{ij} is replaced by the spare PE p_{i0} if $(p_{ij}, p_{i0}) \in M$, and by the spare PE p_{0j} if $(p_{ij}, p_{0j}) \in M$.* □

– Notation: Let $G = (V, E)$ be a bipartite graph where $V = V_1 \cup V_2$ and $V_1 \cap V_2 = \phi$[1]. For $S(\subseteq V_1)$, let $\psi(S) = \{v \in V_2 | (w, v) \in E, w \in S\}$[2]. The degree of a vertex u which is the number of edges incident to u is denoted as $deg(u)$.

It is seen that the degree of any faulty vertex in a compensation graph is equal to or less than two. Using the fact, we characterize the repairability condition as follows.

Theorem 1 *(Repairability Theorem). Let $G = (V, E)$ be a bipartite graph such that $V = V_1 \cup V_2$, $V_1 \cap V_2 = \phi$ and $E \subseteq V_1 \times V_2$ where the degree of any vertex in*

---

[1] G corresponds to a compensation graph in which V_1, V_2 and E correspond to a set of faulty PEs, a set of spare PEs and a set of edges implying replacement relation, respectively.

[2] $(w, v) \in E$ means w can be replaced by v.

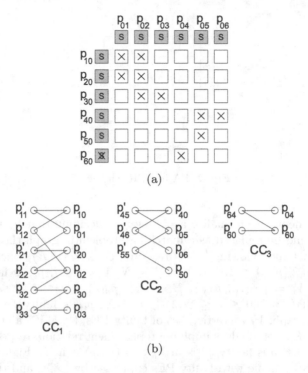

(a)

(b)

Fig. 3. (a) PA in DSR scheme with faulty PEs expressed by "×"s, and (b) it's compensation graph which consists of three connected components CC_1, CC_2 and CC_3.

V_1 is equal to or less than two. We partition the maximal subgraph of G with the vertex set $V_1 \cup \psi(V_1)$ into connected components and denote the vertex sets in V_1 of the connected components as C_1, C_2, ... ,C_m (for each C_p, $C_p \subseteq V_1$, $\psi(C_p)$ $\subseteq V_2$, and for $i \neq j$ $(C_i \cup \psi(C_i)) \cap (C_j \cup \psi(C_j)) = \phi$). Then, the repairability condition is as follows.

> There exists a matching from V_1 to V_2 if and only if $|C_i| \leq |\psi(C_i)|$ for all C_i holds, where $|C|$ means the number of elements of C.

Proof. The only-if-part clearly holds. The if-part: we prove by induction on $|C_i|$. It is sufficient to prove that there exists a matching from C_i to $\psi(C_i)$ for a connected component $G_i = (C_i \cup \psi(C_i), E_i)$ with the vertex set $C_i \cup \psi(C_i)$. The statement of the theorem clearly holds when $|C_i| = 1$. Suppose that for all C_is with $C_i \leq \psi(C_i)$ and $|C_i| \leq m (m \geq 1)$ there exists a matching from C_i to $\psi(C_i)$ and let $m + 1 = |C_i| \leq |\psi(C_i)|$.

(1) The case where there exists a vertex v of degree 1 in $\psi(C_i)$. Since $2 \leq |C_i|$, for $w \in C_i$ such that $(w, v) \in E_i$, $deg(w) = 2$. Let $C_i' = C_i - \{w\}$. Then $\psi(C_i') = \psi(C_i) - \{v\}$ and $G_i' = (C_i' \cup \psi(C_i'), E_i - \{(w, v)\})$ is also connected, and $|\psi(C_i')| = |\psi(C_i)| - 1 \geq m = |C_i'|$. By the hypothesis of induction, there exists a matching M' from C_i' to $\psi(C_i')$. Adding the edge (w, v) to M', a matching from C_i to $\psi(C_i)$ is derived.

(2) The case where the degree of any vertex in $\psi(C_i)$ is equal to or greater than two, that is, there is no vertex of degree 1 in $\psi(C_i)$. Then,

   (i) Since the degree of any vertex in C_i is equal to or less than two, $2 \times |\psi(C_i)| \leq |E_i| \leq 2 \times |C_i|$, which implies that $|C_i| = |\psi(C_i)|$.

 (ii) Suppose that there would exist a vertex of degree 1 in C_i. Then, $2 \times |\psi(C_i)| \leq E_i < 2 \times |C_i|$, which implies that $|\psi(C_i)| < |C_i|$. This is a contraction. Therefore, the degrees of all the vertices in C_i and $\psi(C_i)$ are 2.

(iii) From (i) and (ii), a closed circuit is derived in which the vertices in C_i and $\psi(C_i)$ appear alternatively. This implies that there exist just two different matchings. (In case of a compensation graph, it can easily be shown that $|C_i|(= |\psi(C_i)|)$ for such a closed circuit is even and greater than or equal to 4.)

From (1) and (2), the theorem is proved.   □

According to Theorem 1, we can judge whether a PA with faults is repairable. Furthermore, according to the process proving the theorem, we can decide that a faulty PE is replaced by which spare PE. In the following, Theorem 1 is expressed in a convenient form for realizing the process of finding a matching in hardware, which is given as an algorithm MACH.

Lemma 2. *For any v of degree 1 in V_2 and $(w, v) \in E$, let G' be the graph obtained by removing $\{w, v\}$ from V and the edges incident to w or v. Then, there exists a matching from V_1 to V_2 in G if and only if $V_1 - \{w\}$ to $V_2 - \{v\}$ in G'.*

Proof. From (1) in the proof of Theorem 1.   □

Lemma 3. *For any connected component C_i, let the degree of any vertex in $\psi(C_i)$ be equal to or greater than two. Then,*

(1) If there is a vertex in $\psi(C_i)$ whose degree is greater than two, there exists no matching from V_1 to V_2.

(2) If the degree of every vertex in in $\psi(C_i)$ is two and there exists a vertex of degree 1 in C_i, there exists no matching from V_1 to V_2.

(3) If the degree of any vertex in $\psi(C_i)$ is two and there exists no vertex of degree 1 in C_i, there exists a matching from V_1 to V_2.

Proof. (1), (2) and (3) are proved from the following (1), (2) and (3) with Theorem 1, respectively. (1) since $2|\psi(C_i)| < |E_i| \leq 2|C_i|$, $|\psi(C_i)| < |C_i|$. (2) $|E_i| = 2|\psi(C_i)| < 2|C_i|$. (3) since degree of every vertex in C_i is two, $2|C_i| = 2|\psi(C_i)|$.   □

From Lemmas 2 and 3, we have the algorithm MACH for finding a matching as follows, where Steps 2, 4, 5 and 6 correspond to Lemma 2, (1), (2) and (3) of Lemma 3, respectively.

– Algorithm MACH for finding a matching

1. Let $M = \phi$, $E' = E$, $V_1' = V_1$ and $V_2' = V_2$.
2. Do the following while there is a vertex v of degree 1 in V_2'.
   For $(w, v) \in E'$, let $M = M \cup \{(w, v)\}$, $\hat{E} = \{(w, \hat{v}) | (w, \hat{v})$ in E'$\}$, $E' = E' - \hat{E}$, $V_1' = V_1' - \{w\}$ and $V_2' = V_2' - \{v\}$.
3. If $V_1' = \phi$, M is a matching from V_1 to V_2 and go to 2.
4. If there is a vertex in V_2' whose degree is more than 2, there is no matching from V_1 to V_2 and go to 2.
5. If there is a vertex in V_1' whose degree is 1, there is no matching from V_1 to V_2 and go to 2.
6. There is a matching from V_1 to V_2 and there is a closed cycle in each derived connected component CC_i, from which just two different matching in CC_i are derived. Choose one of them which is denoted as M_i. Let $M = M \cup \{\cup_i M_i\}$. Then M is a matching from V_1 to V_2.
7. The algorithm ends.   □

It is clearly seen from Theorem 1 that the fault pattern shown in Fig. 3 is repairable. On the other hand, we apply MACH to the fault pattern to find a matching, that is, to decide that each faulty PE should be replaced by which spare PE. Then, at the end of Step 2, $M = \{(p_{33}', p_{03}), (p_{32}', p_{30}), (p_{46}', p_{06}), (p_{55}', p_{50}), (p_{45}', p_{05}), (p_{64}', p_{04}), (p_{60}', p_{60})\}$ and a closed cycle shown in Fig. 4 is left in CC_1. Steps 3, 4 and 5 are skipped since the conditions of them are not satisfied. Then, it is judged that there is a matching V_1 to V_2 and in Step 6, it is executed to choose a matching from the closed cycle. Two different matchings derived from the closed cycle are $M_1 = \{(p_{11}', p_{01}), (p_{21}', p_{20}), (p_{22}', p_{02}), (p_{12}', p_{10}),\}$ and $M_2 = \{(p_{11}', p_{10}), (p_{12}', p_{02}), (p_{22}', p_{20}), (p_{21}', p_{01}),\}$. Hence, finally two different matchings from V_1 to V_2 have been derived in Step 6 as $M \cup M_1$ and $M \cup M_2$. By the way, if the labels of the SPs p_{02}, p_{20}, p_{03} and p_{30} are deleted in the closed cycle, the closed cycle with the labels of only the faulty PEs p_{22}', p_{23}', p_{33}' and p_{32}' is derived. In this way, closed cycles obtained in Step 6 in a compensation graph can be in general reduced to closed cycles with only faulty PEs. Such cycles can be drawn directly in a PA, which are searched in hardware as shown in the following section.

In the next section, we realize the procedure of MACH in hardware.

Fig. 4. Closed cycle derived in Step 6 in MACH

3 Hardware Realization

Figure 5 is an example of a mesh array with tracks and switches for replacing faulty non-spare PEs by spare ones. Each PE except spare PEs has four switches around it according to its four I/O ports. Tracks are long links between spare and non-spare PEs. Four tracks run vertically and horizontally between each pair of rows or columns of PEs. If a non-spare PE is faulty, these switch states are changed so that the PE is replaced by a spare PE which is located on the upper or left side of the array.

The algorithm MACH will be realized in hardware for this replacement scheme. In the following, its hardware realization is called "MACH-hard", which consists of two digital circuits. One is NET-1 and another is NET-2. NET-1 outputs a signal whether all the faulty PEs (including spare PEs) are replaced by spare PEs at the same time, while a signal indicating the direction to a spare by which each faulty PE should be replaced is output. These correspond to Steps 2 to 5 in MACH. NET-2 determines the direction to a spare by which each faulty PE in the closed cycles obtained in Step 6 of MACH should be replaced.

The following is the outline of realizing in hardware the above for an $N \times N$ mesh array where the spare PEs are located on the upper and left sides of the array as shown in Fig. 5.

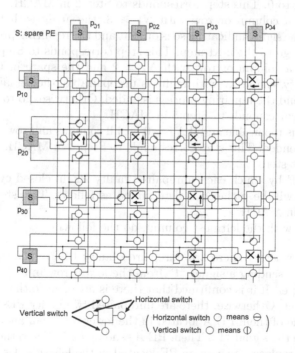

Fig. 5. An example of a mesh array with switches for replacing faulty PEs where arrows indicate directions to spare PEs

Notation:

- The number of unrepaired faulty PEs in a row or column is denoted as N_f.
- An array with faults is said to be repairable (unrepairable) if the set of the faults, that is, the fault pattern is repairable (unrepairable).

Assumption:

- Each of PE and SP outputs 1 as it's fault signal if it is faulty, and 0 otherwise. The fault signal of a PE (SP) is input to the terminal F of MPE (MSP) as shown in Fig. 7.

(Outline of repairing process in hardware realization)

(1) Do the following (a) and (b) N times.
   (a) Count the number of unrepaired faulty PEs (including a spare PE) toward a spare in each column. This is done in parallel for all columns. Then, replace a faulty PE in a column with $N_f = 1$ by a spare in the upper side and set N_f to 0.s
   (b) Count the number of unrepaired faulty PEs (including a spare PE) toward a spare in each row. This is done in parallel for all rows. Then, replace a faulty PE in a row with $N_f = 1$ by a spare in the left side and set N_f to 0. This step corresponds to Step 2 in MACH.
(2) If there is a column or row with $N_f \geq 3$, the array with faults is unrepairable, the signal indicating so is output and the repairing process is ended. Otherwise, go to the next step. This step corresponds to Step 4 in MACH.
(3) If there is a column or row with $N_f = 2$ and the spare in the column or row is faulty, the array with faults is unrepairable, the signal indicating so is output and the repairing process is ended. Otherwise, go to the next step. This step corresponds to Step 5 in MACH.
(4) The array is repairable and if $N_f = 0$ for all columns and rows, the repairing process is ended. This step corresponds to Step 3 in MACH. Otherwise, go to the next step.
(5) A spare PE by which each unrepaired faulty PE in closed cycles in Step 6 in MACH will be replaced is determined as follows. This step corresponds to Step 6 in MACH.
   Beginning with the leftmost column, do the following.

   (i) Check whether there are unrepaired faulty PEs in the column. This is done by sending a signal "1" from the lowest row in the column toward the upper. If it is confirmed that there is none, go to the column next to the right. Otherwise, there are unrepaired two faulty PEs in the column and one of them PE A located in the lower row than another PE B will receive the signal "1". Then, PE A sends signal "1"s to the left and right and is replaced by a spare PE located in the left side. Its general rule is as follows.

- A faulty PE which has received a signal "1" from the upper or lower sends signal "1"s to the right and left, and is replaced by a spare PE located in the left side.
- A faulty PE which has received a signal "1" from the left or right sends signal "1"s to the upper and lower, and is replaced by a spare PE located in the upper side.
- A healthy or repaired PE only passes a signal which it has received.

(ii) Finally, a signal "1" must reach the PE A via PE B. If the column checked is the rightmost column, this process is ended. Otherwise, go to the column next to the right and go to (i).   □

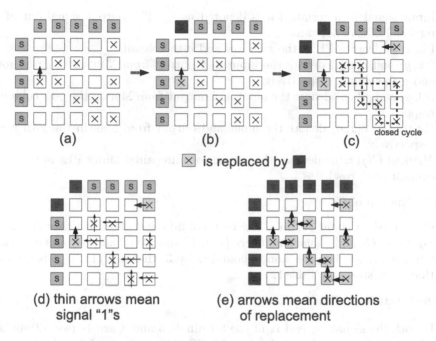

Fig. 6. An example of a repairable fault pattern

Figure 6 illustrates a repairable fault pattern.

In Fig. 6, first, (1) in the outline is applied to (a) and then (c) is derived where the gray squares are the repaired faulty PEs and the black squares are the spare PEs which replaced some faulty PEs. The directions of replacement for two faulty PEs are determined, which are shown by the arrows. Eight faulty PEs are left unrepaired but the conditions of (2), (3) and (4) in the outline are not satisfied. Therefore, this fault pattern is repairable and (5) is applied. The result applied (5) is shown in (d) where the arrows are the signals "1"s sent via the faulty

PEs in the closed cycle $\{p_{32} \to p_{35} \to p_{55} \to p_{54} \to p_{44} \to p_{43} \to p_{23} \to p_{22} \to p_{32}\}$. This leads finally to (e) which shows the directions of all the faulty PEs to be replaced.

Now, first, we show a digital circuit NET-1 which realizes from (2) to (4) in the outline. Next, we show a digital circuit NET-2 which decides the directions of replacements for faulty PEs in closed cycles in (5) in the outline.

Figure 7 shows NET-1 which consists of modules MPE, MSP and a gate G_1 where SP is a spare PE, MPE is the module shown in (b), and MSP is the module shown in (c) which contain submodules C_1s. First, we explain the functions of the modules.

Notation:

- Input signal to a terminal n is denoted as i_n. The output signal out of a terminal m is denoted as o_m.
- $PE(x, y)$ denotes PE in the x-th row and y-th column.
- $i(x, y)$ and $o(x, y)$ denote the input and output from MPE in the x-th row and y-th column, respectively.
- $i(0, y)$ and $o(0, y)$ denote the input and output from MSP in the y-th column, respectively.
- $i(x, 0)$ and $o(x, 0)$ denote the input and output from MSP in the x-th row, respectively.
- $N_f(0, y)$ ($N_f(x, 0)$) denotes the number of unrepaired faulty PEs in the y-th column (x-th row).

• The function of C_1

  C_1 is used to count N_f in a row or a column and check whether $N_f > 2$. To do so, C_1 adds binary numbers $(x_1 x_0)_2$ and $(0f)_2$, and outputs a binary number $(y_1 y_0)_2$ but the sum is bounded by 3, that is, $(11)_2$ if it is greater than 2, as shown in Table 1.

• The function of MPE

1. If both the signals i_5 and i_8 of the terminals 5 and 8 are 1s (so initially as will be shown in 1 of the behavior of NET-1), $o_f = i_F$, and 0 otherwise.
2. If $i_F = 1$ (so if the PE is faulty), the outputs o_U^t and o_L^t of the gate G_2 and G_3 at time t are as shown in Table 2, according to the signals i_5^t and i_8^t of the terminals 5 and 8 at time t. This means that the combination of G_2 and G_3 works as a flip-flop with the inputs i_5 and i_8.

   Note that o_f is input to the terminal F of M_+ in Fig. 8. Further, o_L or o_U is used to indicate the direction to which a faulty PE is replaced by a spare, that is, o_L (o_U) = 1 implies a left (upper) direction.

• The function of MSP

1. If $o_5 = 1$, that is, the output of FF is 1, the output $y_1 y_0$ of C_1 becomes as shown in Table 1.
2. When a clock is input to FF through CK-U (CK-L), if $N_f \leq 1$ in a column (row), Q of FF becomes 0, i.e., $o_5 = 0$, otherwise, 1, i.e., $o_5 = 1$.

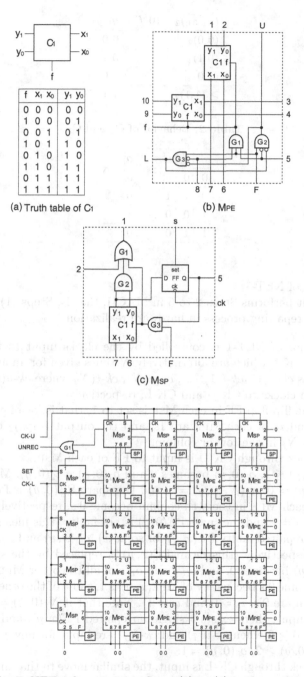

(a) Truth table of C₁

(b) MPE

(c) MSP

Fig. 7. NET-1 for executing Steps (1) to (4) in the outline.

Table 1. Behavior of C_1

$(x_1\, x_0)_2 + (0\, f)_2$	$y_1\, y_0$
$(0\, 0)_2$	0 0
$(0\, 1)_2$	0 1
$(1\, 0)_2$	1 0
$\geq (1\, 1)_2$	1 1

Table 2. Behavior of G_2 and G_3

i_5^t	i_8^t	o_U^t	o_L^t
0	0	$o_U^{(t-1)}$	$o_L^{(t-1)}$
0	1	1	0
1	0	0	1
1	1	0	0

[The behavior of NET-1]

This circuit performs Steps 2 to 5 in MACH, that is, Steps (1) to (4) in the outline of the repairing process in hardware realization.

1. The behavior of NET-1 is controlled by the clocks input to the terminals CK-U and CK-L which are alternatively N times given for an array with size of $N \times N$ as $ck\text{-}u(1)$, $ck\text{-}\ell(1)$, ... , $ck\text{-}u(N)$, $ck\text{-}\ell(N)$ where $ck\text{-}u(i)$ and $ck\text{-}\ell(i)$ are the i-th clocks to CK-U and CK-L, respectively.
2. Initially, the flip-flop FF in each MSP is set to 1, that is, o_5 of each MSP is 1. Then, i_5 and i_8 of each MPE are 1s, and the output $(y_1 y_0)_2$ of C_1 in each MSP shows N_f in the row or column.
3. When a clock through CK-U is input to FF of each MSP,
    (i) if $N_f(0, y) \leq 1$, $o_5(0, y)$ becomes 0 and $o_1(0, y) = i_2(0, y)$. Moreover,
    • If $N_f(0, y) = 0$ or $(N_f(0, y) = 1$ and the spare PE$(0, y)$ is faulty), o_U, o_L and o_f of each MPE in the y-th column become 0's, respectively.
    • If $N_f(0, y) = 1$ and the spare PE$(0, y)$ is not faulty (this means there exists a single unrepaired faulty PE(x, y) for some x), $o_U(x, y)$ and $o_f(x, y)$ become 1 and 0, respectively (this mean PE(x, y) is replaced by the spare PE$(0, y)$ and $N_f(0, y)$ becomes 0, and o_U's and o_L's of all the other MPE's in the y-th column become 0's because i_F of MPE(x, y) is 1, those of the other MPEs in the y-th column are 0's, $i_5 = 0$ and $i_8 = 1$). Note that if $N_f(0, y) = 1$ just before $ck\text{-}u(i)$ is input, there exists a single unrepaired PE(x, y), and $N_f(0, y) \geq 2$ and $N_f(x, 0) \geq 2$ when $ck\text{-}u(j)$ and $ck\text{-}\ell(j)$ are input for any $j < i$.
    (ii) if $N_f(0, y) \geq 2$, $o_5(0, y) = 1$.
4. When a clock through CK-L is input, the similar move to that in the above (3) is performed where "column" is replaced by "row". Note that if $N_f(x, 0) = 1$ just before $ck\text{-}\ell(i)$ is input, there exists a single unrepaired PE(x, y), and $N_f(x, 0) \geq 2$ and $N_f(0, y) \geq 2$ when $ck\text{-}\ell(j)$ and $ck\text{-}u(j)$ are input for any $j < i$.

5. The clocks through CK-U and CK-L are alternatively N times given. A faulty PE whose o_f is 0 has been repaired. After this, such a faulty PE is not counted in N_f because o_f is 0. The above (3) to (5) correspond to (1) in the outline.

6. $o_{\mathrm{UNREC}} = 1$ if and only if o_1 of some MSP is 1 if and only if $N_f > 2$ or ($N_f = 2$ including a faulty spare) in a row or a column. Hence, $o_1 = 1$ of some MSP corresponds to (2) or (3) in the outline and $o_{\mathrm{UNREC}} = 1$ of NET-1 indicates that the array with the faults is unrepairable.

7. If $o_{\mathrm{UNREC}} = 0$, o_1's of all MSP's are 0's, which indicates that the array with the faults is repairable. Then, if there is neither row nor column such that $N_f = 2$, the spare is healthy and o_5 of MSP is 1, this repairing process can be successfully ended. However, for simplicity, omitting this check, we go to the process for finding closed cycles (even if there may not be such cycles) together with the directions of replacing faulty PEs in the cycles by spares. This process corresponds to (5) in the outline and is realized by NET-2 shown in Fig. 8.   □

Next, the behavior of NET-2 shown in Fig. 8 is described.

• The function of M+

1. The terminal f of each MPE in NET-1 is connected to the terminal F of M+ corresponding to the MPE.

2. If i_F ($= o_f$) is 0, the internal structure of M+ becomes as shown in "(b) The case of F = 0" in Fig. 8, i.e., the signals pass through horizontally and vertically.

3. If i_F is 1, the internal structure becomes as shown in "(a) The case of F = 1" in Fig. 8.

   The signal through x_3 from the top or x_1 from the bottom are transferred to the left through y_4 and the right through y_2, and the signal is stored in the flip-flop FFL, which indicates that the direction of replacement is to the left if the signal is 1.

   The signal through x_4 from the left or x_2 from the right are transferred to the lower through y_1 and the upper through y_3, and the signal is stored in the flip-flop FFU, which indicates that the direction of replacement is to the upper if the signal is 1.

[The behavior of NET-2]

This circuit performs Step 6 in MACH, that is, (5) in the outline of the repairing process. First, Note that the internal structure of M+ becomes as shown in "(b) The case of F = 0" in Fig. 8 if a PE with the M+ is healthy or has been repaired and in "(a) The case of F = 1" if it is faulty and has not yet been repaired. Further, note that there are exactly two unrepaired faulty PEs in a row or column in a closed cycle.

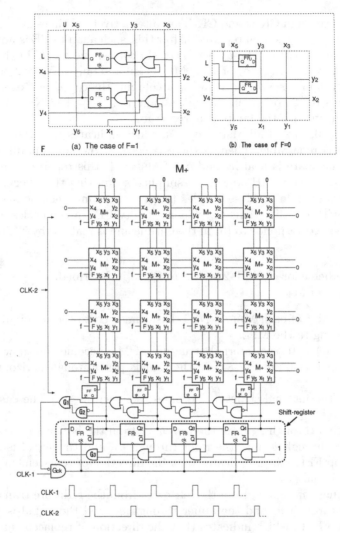

Fig. 8. NET-2 for deciding the directions of replacement while executing Step (5) in the outline

1. Initially, all the flip-flops are reset.
2. Signal 1 is shifted from the left to the right in the shift-register at the time when a clock pulse is given to CLK-1.
3. For $i = 1$ to N, the following is performed.
   (i) A clock is fed to all the flip-flops in Fig. 8 through CLK-2 except ones in the shift-register. Q_i of the shift-register becomes 1. This signal "1" is input to x_1 of M_+ in the bottom row of i-th column. At the time, the output of the gate G_1 becomes 1 and hence a clock to CLK-1 is inhibited to be supplied to the shift register. While a clock to CLK-1 is not supplied, the signal "1" input to the i-th column behaves as follow.

- If there is no unrepaired faulty PE in the i-th column, the signal "1" passes through all the M_+s in the column, turns back at the M_+ in the top row of the column (note that the terminals x_5 and y_3 are connected) and reaches the terminal y_5 of the M_+ in the bottom row of the column.
- If there are unrepaired faulty PEs, there are exactly two such PEs in the column whose M_+s are denoted as M_+^L and M_+^U where the former is in a lower location. The signal "1" is fed to M_+^L and propagates in a closed cycle as mentioned in (5) in the outline, finally reaches M_+^U, sent in the upper direction, turn back at M_+ in the top row and reaches the terminal y_5 of the M_+ in the bottom row of the column.
- The signal "1" which reaches y_5 of the M_+ in the bottom row as above is memorized in the D-FF by a clock to CLK-2 and fed to the gate G_2. Then, the outputs of G_2 and G_1 become 0s, and hence, a clock to CLK-1 can pass through the gate G_{ck}. □

From explained so far, it is seen that NET-1 and NET-2 exactly execute each step in MACH. That is, a fault pattern satisfies the repairability condition in Theorem 1 if and only it is repaired using NET-1 and NET-2. This means that the selection of routes for replacement is done in an optimal way in the meaning of satisfying the repairability condition.

4   Simulations and Results

To estimate the effects of faults to the array reliability, considering not only faults in PEs but also in the MACH-hard. It is assumed that tracks and switches are fault-free because they are very simple compared to PEs and the MACH-hard in common cases. We execute Monte Carlo simulations. We use a PC/AT machine with 2.2 GHz Celeron CPU, 1 GB DDR2 RAM, Fedora 14 Linux OS, and gcc 4.5.1 C compiler. In the simulation program, faulty PE locations are specified using the C language standard function "random()" which is a pseudo-random number generator.

The main part of the simulation program is the MACH and exhaustive algorithms for DSR and 1.5-TS schemes, respectively, where the exhaustive algorithm exhaustively tries to find a matching for array reconstruct. The data are calculated by the following equations.

Let $N_{tested}(k)$ be the number of sets of faulty PE indices with k faulty PEs which are tested by the algorithm. Let $N_{reconfigurable}(k)$ be the number of them which are judged reconfigurable by the algorithm in $N_{tested}(k)$ tested sets. In the simulation program, $N_{tested}(k)$ is set to 500 for each k. Let

$$AR_{alg}(p) = \sum_{k=0}^{N_s} {}_{N_a}C_k \cdot AY(k) \cdot p^{N_a-k} \cdot (1-p)^k \tag{1}$$

where p is the PE reliability, N_s is the number of spare PEs, N_a is the number of PEs, ${}_iC_j = \frac{i \cdot (i-1) \cdot (i-2) \cdots (i-j+1)}{j \cdot (j-1) \cdot (j-2) \cdots 1}$, and $AY(k) = N_{reconfigurable}(k)/N_{tested}(k)$.

The Eq. (1) means the average value of $AY(k)$ which is calculated using the probability that just k PEs are faulty, that is, $_{N_a}C_k \cdot p^{N_a-k} \cdot (1-p)^k$. The Eq. (1) also means the array reliability of the array with tracks and switches in Fig. 5 being fault-free. Note that the data in Fig. 1 are calculated using the Eq. (1). Let

$$AR_{mh}(p) = AR_{alg}(p) \cdot p_{mh}^{N_{ns}}$$
$$= AR_{alg}(p) \cdot p^{\frac{N_{ns}}{M_{mh}}} \qquad (2)$$

where p_{mh} is the reliability of the MACH-hard per non-spare PE, N_{ns} is the number of non-spare PEs, and $M_{mh} = \frac{log(p)}{log(p_{mh})}$, that is, the ratio of the hardware complexity of a PE to that of the MACH-hard per non-spare PE $(p = p_{mh}^{M_{mh}})$. The Eq. (2) means the array reliability considering not only faults in PEs but also in the MACH-hard because $p_{mh}^{N_{ns}}$ means the reliability of the whole MACH-hard.

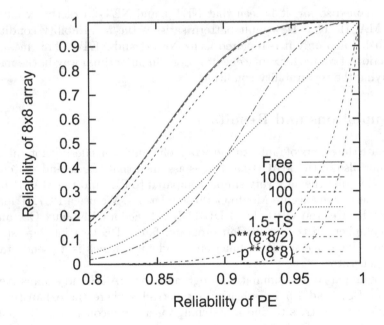

Fig. 9. The relation between AR_{mh} and p (8×8)

Figures 9, 10 and 11 show the relations between AR_{mh} and p for 8×8, 16×16 and 32×32 arrays, respectively. The label "Free" indicates AR_{alg} for a DSR array, that is, the array reliability where tracks, switches and the MACH-hard are fault-free. The label of the integer indicates the value of M_{mh} for a DSR array. The labels "p**(N*N)" and "p**(N*N/2)", where N is 8, 16 or 32, indicate $p^{N \times N}$ and $p^{\frac{N \times N}{2}}$, respectively.

Here, the method to realize the fault signal of each PE is considered. The technique that the function of a PE in a non-redundant array is duplicated and

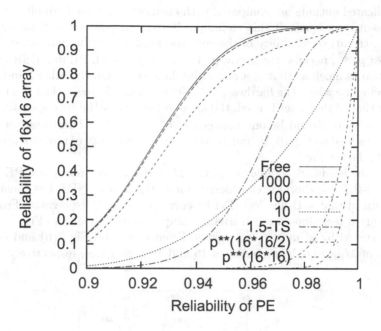

Fig. 10. The relation between AR_{mh} and p (16×16)

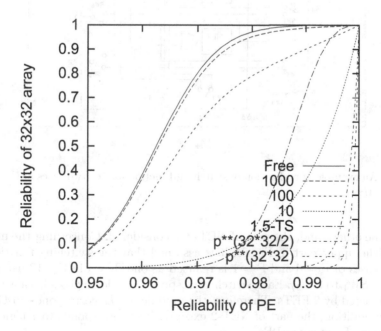

Fig. 11. The relation between AR_{mh} and p (32×32)

the duplicated outputs are compared with each other is called "duplication and comparison technique". If this technique is used to realize the fault signal of each PE, the array reliability of the corresponding non-redundant array should be almost $p^{\frac{N\times N}{2}}$, because the hardware complexity of each PE in a DSR array is almost two as much as that in a non-redundant array. On the other hand, if the other technique where the hardware complexity of the former is less than two as much as that of the latter is used, the array reliability of the corresponding non-redundant array should be one between $p^{\frac{N\times N}{2}}$ and $p^{N\times N}$. Therefore, these two curves are added to Figs. 9, 10 and 11 as the array reliability of the corresponding non-redundant array.

Here, the hardware complexity of the MACH-hard per non-spare PE ("HC-MACH-ns-PE" for short) is considered. First, the part of NET-1 is considered. It is assumed that a D-flip-flop (D-FF) corresponds to 2 logic gates. From the simplification technique using Karnaugh map, the module of C_1 (Fig. 7(a)) can be constructed by at most 6 logic gates. From these and Fig. 7(b) and (c), the modules of M_{PE} and M_{SP} consist of 15 and 10 logic gates, respectively.

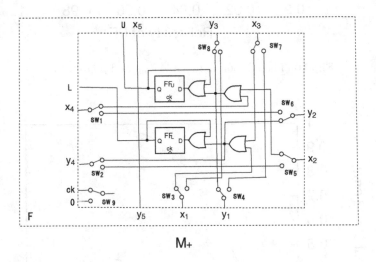

Fig. 12. An example of realization of switching function for the two cases of F = 0 and F = 1 of M_+

On the other hand, the part of NET-2 is considered. Concerning the module of M_+ (the upper part in Fig. 8), it is assumed that the switching function for the two cases of F = 0 and F = 1 is realized as one shown in Fig. 12 using nine switches (SW_1 to SW_9). Each switch corresponds to 1 logic gates because it can be constructed by 2 FETs. Therefore, the module of M_+ corresponds to 17 logic gates. In addition, the part of NET-2 except M_+s[3] corresponds to 7 logic gates per column of non-spare PEs.

---

[3] This is the bottom part in Fig. 8 including the part of "Shift-register".

Table 3. HC-MACH-ns-PE (logic gates)

array size	HC-MACH-ns-PE
8×8	35.4
16×16	33.7
32×32	32.8

From the above, the value of HC-MACH-ns-PE (logic gates) can be calculated for each case of array size and Table 3 shows these calculated values.

From Figs. 9, 10 and 11 and Table 3, we can see the following.

- The faults in the MACH-Hard in the case that $M_{mh} > 1000$ are negligible because in Figs. 9, 10 and 11 the curves with the label "1000" are almost the same as their corresponding curves with the label "Free", respectively. The case that $M_{mh} > 1000$ is generally a common case because the case means that the hardware complexity of a PE is greater than 35400 logic gates and the hardware complexities of most general purpose CPUs for PCs and even micro-controllers are so.
- The curves with the label "100" are greater for most p and a little less for some p than their corresponding curves with the label "1.5-TS", respectively. This shows the advantage of DSR scheme in terms of array reliability compared to 1.5-TS scheme in the same PE redundancy.
- The curves with the label "10" are greater than their corresponding curves with the labels "p**(N*N)" and "p**(N*N/2)", respectively. This shows the usefulness of DSR scheme to design redundant mesh-connected PAs.

5 Conclusion

We have presented a built-in digital circuit for making mesh-connected arrays fault-tolerant by replacing faulty PEs directly by spare PEs located at two orthogonal sides of the arrays. Mazumder et al. [13] developed a circuit using an analog neural network for this scheme. However, it not only restructures arrays in a heuristic way in selecting routes for replacing faulty PEs by spares, but also has some weak points in hardware realization as mentioned in Introduction. On the other hand, in this paper, a digital circuit more reliable than an analog one is used and the selection of routes for replacement is done in an optimal way. The circuit consists of arrays of simple modules and can easily be embedded in a target PA to recover from the fault without the aid of a host computer. This implies that the proposed system is so useful in not only enhancing especially the run-time reliabilities of PAs but also such an environment that the repair by hand is difficult or a PA is embedded within a VLSI chip where the fault pattern cannot be monitored externally through the boundary pins of the chip, and so on. In addition, the data of the array reliability from Monte Carlo simulations, considering not only faults in PEs but also in the MACH-Hard, are given, and then the effectiveness of the DSR scheme is shown.

Detail analysis for overheads of area and power consumption, and time delay between PEs and spares is a future work.

References

1. Kung, S.Y., Jean, S.N., Chang, C.W.: Fault-tolerant array processors using single-track switches. IEEE Trans. Comput. **38**(4), 501–514 (1989)
2. Mangir, T.E., Avizienis, A.: Fault-tolerant design for VLSI: Effect of interconnection requirements on yield improvement of VLSI designs. IEEE Trans. Comput. c–**31**(7), 609–615 (1982)
3. Leighton, T., Leiserson, E.: Wafer-scale integration of systolic arrays. IEEE Trans. Comput. **C–34**(5), 448–461 (1985)
4. Lam, C.W.H., Li, H.F., Jakakumar, R.: A study of two approaches for reconfiguring fault-tolerant systolic arrays. IEEE Trans. Comput. **38**(6), 833–844 (1989)
5. Kim, J.H., Reddy, S.M.: On the design of fault-tolerant two-dimensional systolic arrays for yield enhancement. IEEE Trans. Comput. **38**(4), 515–525 (1989)
6. Negrini, R., Sami, M.G., Stefanelli, R.: Fault-tolerance through reconfiguration of VLSI and WSI arrays. MIT Press series in computer systems. MIT Press, Cambridge (1989)
7. Koren, I., Singh, A.D.: Fault tolerance in VLSI circuits. IEEE Comput. **23**, 73–83 (1990)
8. Dutt, S., Hayers, J.P.: Some practical issues in the design of fault-tolerant multiprocessors. IEEE Trans. Comput. **41**(5), 588–598 (1992)
9. Shigei, N., Miyajima, H., Murashima, S.: On efficient spare arrangements and an algorithm with relocating spares for reconfiguring processor arrays. IEICE Trans. Fundam. **E80–A**(6), 988–995 (1997)
10. Roychowdhury, V.P., Bruck, J., Kailath, T.: Efficient algorithms for reconstruction in VLSI/WSI array. IEEE Trans. Comput. **39**(4), 480–489 (1989)
11. Sami, M., Negrini, R., Stefanelli, R.: Fault tolerance techniues for array structures used in supercomputing. IEEE Comput. **19**(2), 78–87 (1986)
12. Sami, M., Stefanelli, R.: Reconfigurable architectures for VLSI processing arrays. Proc. IEEE **74**, 712–722 (1986)
13. Mazumder, P., Jih, Y.S.: Restructuring of square processor arrays by built-in self-repair circuit. IEEE Trans. Comput. Aided Des. **12**(9), 1255–1265 (1993)
14. Takanami, I., Kurata, K., Watanabe, T.: A neural algorithm for reconstructing mesh-connected processor arrays using single-track switches. Int. Conf. on WSI, 101–110 (1995)
15. Horita, T., Takanami, I.: An efficient method for reconfiguring the $1\frac{1}{2}$ track-switch mesh array. IEICE Trans. Inf. Syst. **E82–D**(12), 1545–1553 (1999)
16. Horita, T., Takanami, I.: Fault tolerant processor arrays based on the $1\frac{1}{2}$-track switches with flexible spare distributions. IEEE Trans. Comput. **49**(6), 542–552 (2000)
17. Horita, T., Takanami, I.: An FPGA implementation of a self-reconfigurable system for the $1\frac{1}{2}$ track-switch 2-D mesh array with PE faults. IEICE Trans. Inf. Syst. **E83–D**(8), 1701–1705 (2000)
18. Horita, T., Takanami, I.: A system for efficiently self-reconstructing $1\frac{1}{2}$-track switch torus arrays. IEICE Trans. Inf. Syst. **E84–D**(12), 1801–1809 (2001)
19. Horita, T., Takanami, I.: An efficiently self-reconstructing array system using E-$1\frac{1}{2}$-track switches. IEICE Trans. Inf. Syst. **E86–D**(12), 2743–2752 (2003)
20. Lin, S.Y., Shen, W.C., Hsu, C.C., Wu, A.Y.: Fault-tolerant router with built-in self-test/self-diagnosis and fault-isolation circuits for 2D-mesh based chip multiprocessor systems. Int. J. Electr. Eng. **16**(3), 213–222 (2009)

21. Collet, J.H., Zajac, P., Psarakis, M., Gizopoulos, D.: Chip self-organization and fault-tolerance in massively defective multicore arrays. IEEE Trans. Dependable Secure Comput. **8**(2), 207–217 (2011)
22. Takanami, I.: Self-reconfiguring of $1\frac{1}{2}$-track-switch mesh arrays with spares on one row and one column by simple built-in circuit. IEICE Trans. Inf. Syst. **87**(10), 2318–2328 (2004)
23. Melhem, R.G.: Bi-level reconfigurations of fault tolerant arrays. IEEE Trans. Comput. **41**(2), 231–239 (1992)
24. Sugihara, K., Kikuno, T.: Analysis of fault tolerance of reconfigurable arrays using spare processors. IEICE Trans. Inf. Syst. **E75–D**(3), 315–324 (1992)
25. Takanami, I., Horita, T.: A built-in circuit for self-repairing mesh-connected processor arrays by direct spare replacement. IEEE Int. Symp. on PRDC, 96–104 (2012)

A Secure Encryption Method for Biometric Templates Based on Chaotic Theory

Garima Mehta[1]([⊠]), Malay Kishore Dutta[1], and Pyung Soo Kim[2]

[1] Department of Computer Science and Engineering,
Amity University, Noida, Uttar Pradesh, India
{gmehta,mkdutta}@amity.edu
[2] Electronic Engineering, Korea Polytechnic University, Siheung, Korea
pskim@kpu.ac.kr

Abstract. This paper presents an encryption based security solution for iris biometric template for secure transmission and database storage. Unlike conventional methods where raw biometric images are encrypted, this paper proposes method for encryption of biometric templates. The advantage of this method is reduced computational complexity as templates are smaller in size than the original biometric image making it suitable for real time applications. To increase the security of the biometric template, encryption is done by using the concept of multiple 1-D chaos and 2-D Arnold chaotic map. The proposed scheme provides a large key space and a high order of resistance against various attacks. Template matching parameters like hamming distance, weighted Euclidean distance, and normalized correlation coefficient are calculated to evaluate the performance of the encryption technique. The proposed algorithm has good key sensitivity, robustness against statistical and differential attacks and an efficient and lossless method for encrypting biometric templates.

Keywords: Biometric template security · Encryption · Chaotic system · Arnold cat map · Logistic map

1 Introduction

With extensive deployment of biometric authentication systems in public and private sector, the need for securing these systems from unauthorized access is highly required. "Biometrics" refers to the distinctive measurable physical or physiological characteristics like fingerprints, iris, face, palm prints, gait, voice, signature, and keystrokes etc. used for recognizing the identity, or authenticating the claimed identity of an individual thereby providing a reliable solution for user authentication [1]. Among these biometrics iris is one of the best because of its stability, uniqueness and non-invasiveness [2]. The main advantage of any biometric over the traditional method [3] is that while verification/authentication of any person, he/she must be physically present at that place as compared to password mechanism system, where the difference between the attacker and an authorized person/user cannot be determined. Hence, we can say that biometric is a strong weapon than any traditional authentication method. However biometric recognition systems are also prone to deliberate attacks as well as inadvertent security

© Springer-Verlag Berlin Heidelberg 2016
M.L. Gavrilova and C.J. Kenneth Tan (Eds.): Trans. on Comput. Sci. XXVII, LNCS 9570, pp. 120–140, 2016.
DOI: 10.1007/978-3-662-50412-3_8

lapses that can lead to illegitimate intrusion [4, 20], sabotage or theft of sensitive information such as the biometric templates of the users enrolled in the system, so there is a need to secure these biometric templates by combining it with cryptography over transmission channels and in storage databases.

There are certain issues [5] regarding the privacy of biometric template like:

- Biometrics is not secret: An attacker can make/duplicate a biometric image such as fingerprint, face, iris, etc without the knowledge of the actual owner of the biometric.
- Biometric cannot be revoked: Any individual's biometrics is permanently associated with him/her. It is difficult to revoke it in the event of any fraud.
- Biometrics can be utilized in the multiple applications as they remain the same for an individual.

Conventional cryptography techniques like Advance Encryption Standard (AES), Rivest–Shamir–Adleman (RSA) and Data Encryption Standard (DES) [25] are not suitable for biometric templates due to inseparable characteristics of biometric data like high correlation among adjacent pixels, high redundancy etc. [6]. This leads to the need to develop different encryption algorithms to solve these issues of biometric templates. Chaotic theory is one of the best suited methods for use in encryption techniques for biometric templates as biometric templates show chaotic behavior in spatial domain [7, 21] which makes it difficult for an intruder to access the data under attacks because of the high key sensitivity, pseudorandom nature and high robustness towards attacks.

The main contribution of the paper is a method where biometric template extracted from an iris image is encrypted for secure database storage or transmission over unsecured data channels. The template of the biometric image is used for unique identification and hence encrypting the template will reduce the computational and space complexity in comparison to the encryption of the image. Chaotic theory is preferred for encryption of biometric template as it is highly sensitive to initial conditions, pseudorandom in nature, and has high resistivity against attacks [13]. The proposed encryption scheme uses concept of multiple 1-D chaos and 2-D Arnold chaotic map for substitution and shuffling respectively to enhance the security of biometric template. The proposed chaos based encryption algorithm for securing biometric templates is an efficient, lossless and robust method having high key sensitivity. Unlike previous reported work, this unconventional method to encrypt the templates instead of the image will suit the real time applications.

The paper is organized as follows. Section 2 describes the related work. The next section describes the proposed methodology. Section 4 presents the experimental results. Finally, Sect. 5 concludes the paper.

2 Related Work

Most of the work reported in the area of biometric encryption [10, 11] is done on biometric images. Encryption of biometric images may not be recommended good choice for real time applications because of its size and large computational time will be involved. Even though biometric systems supports huge databases, large size image storage is a serious issue as it reduces the performance of the biometric system.

In most of the existing work [14–16] encryption of biometric images like finger-print, iris, palm print is done to enhance the transmission security, using chaotic, fractional and combinational domains. Additional work also has been done [15] for securing the individual identification. But in all these previous reported work of bio-metric security the biometric images are used either for transmission over insecure network channels or for securing the individual identification. The biometric systems which are fast gaining importance for security purposes or for recognition/ authenti-cation use the concept of storing the biometric template rather than biometric images due to large size, bulk data capacity, large computational time and large amount of processing power for biometric images. In addition all the parts of the biometric image are not useful, thus it leads to wastage of storage capacity, additional computational time etc. Though, biometric systems have large databases storing thousands of data of different biometric users, the storing images of large size may not be possible. Therefore only the unique features are extracted [12] from these biometric images and are stored as a template of the image in the database. To enhance the transmission security, online authentication and for database storage, encryption of template seems to be more preferred proposition over encryption of biometric images.

Biometric templates may not be secure from malicious attacks by unauthorized users [5]. Thus to restrict these illegal activities by unauthorized parties security of biometric template is an important requirement. Transferring of biometric template over insecure network channels [17] and storage of biometric template in a database both are the most vulnerable points of attack from where the data may be accessed by invaders. The attack on biometric template stored in the database may lead to several vulnerabilities like a template can be replaced by an impostor's template to gain unauthorized access. A physical spoof can be created from the original one to gain unauthorized access to the system. Thus to encounter these vulnerabilities, the concept of combining biometrics with cryptography may be a potential solution to such threats. Instead of storing the biometric template in raw form in the database or to transmit biometric template over communication channels from one end to another in raw form, encryption is an effective way to prevent the leakage of information and to withstand malicious attacks by unauthorized users.

3 Proposed Methodology

This section describes the encryption method for biometric iris image in detail. The first step towards proposed algorithm is feature extraction and template generation which will be encrypted for security. In this proposed work, the shuffling and substitution architecture is followed where shuffling/permutation is done by Arnold chaotic map and combination of two logistic maps are used as a pseudorandom number generator for substituting the pixel values which cause a large variation in output with small variation in input making chaotic system different from classical encryption systems.

The unique characteristics of this combination are good sensitivity towards initial conditions, non periodic and non-convergence. Large key space is provided by this chaotic encryption method which makes it resistant to brute force attacks. Small deviation in set of initial values causes drastic change in output thereby, making the proposed algorithm effective, robust and efficient. The proposed algorithm uses the biometric template as an input for encryption and once encrypted, the template can be used for transmission or storage.

Proposed Algorithm

(i)   Iris image is scanned
(ii) Feature extraction of iris image is done to generate a digital pattern called biometric template using Daugmann's rubber sheet model.
(iii) Zero Padding is done to convert rectangular template of size 20*480 to square sized template of 100*100.
(iv) Apply Arnold Transformation for generating pixel value mixing effect.
(v)   Generate substitution matrix by using combination of two logistic maps.
(vi) Using substitution matrix and pixel mixed matrix the biometric template is encrypted.
(vii) Biometric Template is decrypted at receiver end using a correct combination of key.

3.1   Feature Extraction from Iris Image and Template Generation

Feature extraction is the process where biometric image obtained from the scanner is processed to generate a pattern known as biometric template. A modified Daugman's rubber sheet model [9] is used for extraction of template from the iris image. One reason behind choosing Daugman's rubber sheet model is its popularity and high rate of acceptance among the biometric fraternity. It is used in this work because in this method the angular and radial resolution can be controlled and maintain sufficient information for uniqueness. Feature extraction process from an iris sample may be explained in three steps:

Segmentation. The initial phase of iris image processing is the *segmentation* phase which is used to separate the iris's spatial extent in the eye image by isolating it from the other structures of complete digital eye image such as the sclera, pupil, eyelids, and eyelashes [9].

Normalization. Once the iris boundary has been estimated by segmentation module a method based on Daugman's rubber sheet model [8, 9] was used for normalization of iris images in which each point in the iris region is transformed to pair of polar coordinates (r, θ) where r is in interval [0, 1] and θ is the angle $[0, 2\pi]$.

Feature Encoding. Feature Encoding is done by convolving the normalized eye pattern with 1-D Log-Gabor wavelets.1-D signals are formed by breaking up of 2-D normalized eye pattern and then these 1-D signals are convolved with 1D Gabor wavelets [8]. Figure 1 shows the feature extraction and template generation procedure.

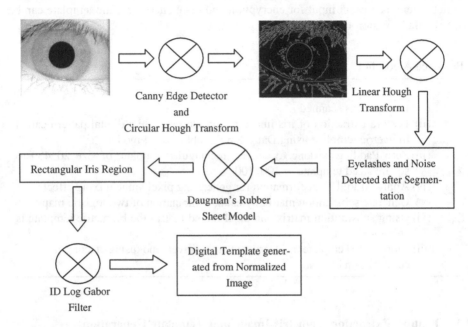

Fig. 1. Feature extraction and template generation from an iris image

The template generated after feature encoding phase using Gabor filter is of size 20*480 which is further resized into square sized template of size 100*100 by zero padding at the end of the template for the encryption process. The square sized template is required for Arnold transformation.

3.2   Application of Arnold Chaotic Map in Biometric Template Encryption

Arnold cat map is a two dimensional invertible chaotic map which is used for shuffling and wrapping operation among the pixel positions of biometric template to generate a scrambled matrix after several iterations. In Arnold cat map a square matrix is stretched by performing transformation followed by a folding operation by modulo function to make the cat map area preserving.

Mathematically, Arnold cat map [19] is described as

$$\begin{bmatrix} x_{n+1} \\ y_{n+1} \end{bmatrix} = \begin{bmatrix} a & b \\ c & d \end{bmatrix} \begin{bmatrix} x_n \\ y_n \end{bmatrix} \mod M \qquad (1)$$

where a, b, c and d are positive integers, ad-by = $\pm$ 1.
x_n, y_n, x_{n+1} and y_{n+1} are integers in {0, 1, 2,.., M-1}.

In case only positive integers are used to generate values, the cat map becomes periodic in nature. The (x_n, y_n) is the pixel position in the original biometric template, and the (x_{n+1}, y_{n+1}) is the new pixel position of the pixel (x_n, y_n). Since ad-bc = $\pm$ 1 the above expression can be modified as:

$$\begin{bmatrix} x_{n+1} \\ y_{n+1} \end{bmatrix} = \begin{bmatrix} 1 & e \\ f & ef+1 \end{bmatrix} \begin{bmatrix} x_n \\ y_n \end{bmatrix} \mod M \qquad (2)$$

T is the periodicity of the image of size M*M which depends on the parameters e, f and size M of the original image. The parameter e, f and iterating times N can be used as the encryption keys of the encryption algorithm. Following Table 1 shows the value of T for different M*M images.

Table 1. Period table for arnold cat map

Period table					
Size	*50*	*74*	*100*	*150*	*256*
Period	150	114	150	300	192

In this paper, Arnold cat map is used to perform scrambling of pixels to encrypt the biometric template. A 20*480 sized biometric template extracted from iris image from feature extraction method is converted into a 100*100 square sized template to apply Arnold cat map. The periodicity of the image can be calculated using Eq. 2. Here the parameters e, f is taken as 1 and as the size of template is 100*100, calculation shows that the periodicity is equal to 150 which is used in the proposed method.

3.3 Application of 1-D Logistic Map in Biometric Template Encryption

Logistic map (http://mathworld.wolfram.com/Logistic Map.html) is the one dimensional polynomial chaotic map which generates a sequence of real numbers and can be expressed as

$$x_{k+1} = rx_k(1 - x_k) \qquad (3)$$

where x_k can take the value between 0 and 1 and represents the population at year k and x_0 represents the initial population whereas r is a positive number that can take value between 0 and 4 and shows the combined rate for reproduction and starvation. For the value of r between 3.57 and 4 the equation shows the chaotic nature and obtained sequence shows non periodic and non convergent behavior and also show high sensitivity towards initial conditions because at r=3.57 no longer oscillations can be seen.

In this paper, logistic map is used to generate pseudorandom numbers which are used to substitute the original values. For this purpose two different logistic maps are used to obtain two different sequences of pseudorandom numbers and a competitive sequence is obtained to increase the key structure complexity of the proposed algorithm. Bifurcation diagram and Lyapunov exponent are the comparative factors for evaluating different chaotic maps. The positive value of Lyapunov exponent determines the chaotic behavior. Figure 2(a) and (b) shows the bifurcation diagram and Lyapunov exponent of logistic map.

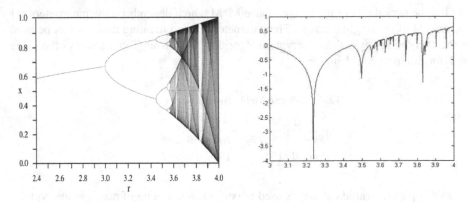

Fig. 2. **(a) and (b)** Bifurcation diagram of logistic map and Lyapunov exponent

Bifurcation diagram of logistic map displays the qualitative information about equilibria of the logistic map. It is obtained by plotting r for a series of value of x_n. The bifurcation parameter r is shown on horizontal axis of the plot diagram and the vertical axis shows the possible long-term population values of the logistic function. The Lyapunov exponent for a map is expressed as follows:

$$\lambda = \frac{1}{n}\sum_{i=0}^{n-1} \log_e |C'[x_i; r]| \tag{4}$$

Figure 2(b) shows the Lyapunov exponent of the logistic map and it is observed that in logistic map for some values of r, Lyapunov constant is either zero or negative in chaotic region [10].

3.4   Proposed Encryption Algorithm

The proposed encryption technique for biometric template is described as follows:

Encryption Process

(i)   Perform n_1 iterations of Arnold Cat Map on reshaped template of size 100*100 to generate a semi-encrypted template SET_1.

(ii) Based on 2 different logistic maps generate two different pseudo-random sequence of numbers and using competition between two, generate a 100*100 substitution matrix S_1.

(iii) The above generated substitution matrix S_1 is bitxor with semi-encrypted template SET_1.

(iv) The output of above step generates the finally encrypted template ET_1.

(v)   The encrypted template ET_1 is stored in the Template Database.

(vi) The encryption key EK is stored separately to be used in the decryption process.

For decryption, either at the receiver end or for biometric authentication, the encryption key EK is required. The decryption process is as follows:

Decryption Process

(i)   Read the encrypted template from the database.

(ii) Based on 2 different logistic maps generate two different pseudo-random sequence of numbers and using competition between two, generate a 100*100 substitution matrix S1.

(iii) The above generated substitution matrix S1 is subjected to bitxor with encrypted template to produce the semi-decrypted template SET2.

(iv) This SET2 is then subjected to n2 i.e. (y-n1) iterations of Arnold Cat map to obtain the decrypted template.

3.5   Block Diagram for Proposed Scheme

In this section proposed encryption technique is illustrated with the help of block diagram in Fig. 3.

Figure 3 presents the block diagram for the proposed encryption process. The steps involved in encryption process and decryption process are as follows:

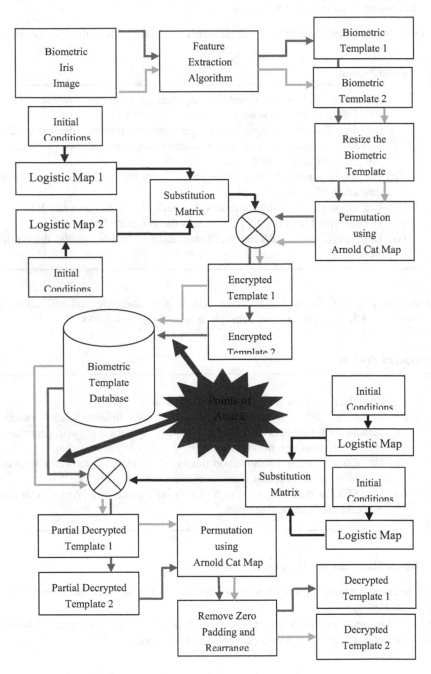

Fig. 3. Block diagram of proposed encryption process

4   Experimental Results

In this section the experimental results for template matching parameters, robustness against statistical attacks and differential attacks for the proposed algorithm have been discussed in Table 2. Sample iris images are used for experiments from CASIA database. All iris images are 8 bit gray-level JPEG files, collected under near infrared illumination.

The following Table 2 shows the iris image and its corresponding processed elements i.e. extracted and reshaped template, encrypted template and decrypted template.

4.1   Template Matching Parameters

The proposed algorithm has been evaluated for the template matching parameters like hamming distance, normalized correlation coefficient and weighted Euclidean distance to check whether the template after decryption is same as that of original one.

Hamming Distance. Hamming distance is a metric used for template matching which is used to find the number of similar bits between original and encrypted biometric template [8]. Hamming Distance [8] is calculated using the following formula:

$$HD = \frac{1}{N} \sum_{i=1}^{N} X_i (XOR) Y_i \qquad (5)$$

where N is the total number of bits in a template and X_i represents the bit in the original biometric template and Y_i represents the bit in encrypted biometric template.

The experimentally calculated hamming distance between original and encrypted template and between original and decrypted templates has been shown in Fig. 4.

The hamming distance between original and encrypted template presented in indicates both templates are independent and bits are not correlated as the calculated value is close to 0.5 and there are 0.5 chances of setting 1 bit to zero and vice versa, whereas hamming distance between original and decrypted comes out to be 0 which shows that after decryption template is similar to that of original one and there is no dissimilar bit between original and decrypted template. As per matching criterion required for template matching, the hamming distance between original and encrypted comes out to be 0.5 whereas hamming distance b/w original and decrypted comes out to be zero which indicates efficient encryption. As tabulated in Fig. 4. the experimental results are encouraging and fulfill the conditions laid down for template matching.

Normalized Correlation. Normalized correlation is another metric [8] used for template matching. To evaluate the normalized correlation [8] between original, encrypted and decrypted biometric templates, the following formula has been used.

$$\frac{\sum_{i=1}^{n} \sum_{j=1}^{m} (p_1[i,j] - \mu_1)(p_2[i,j] - \mu_2)}{nm\sigma_1\sigma_2} \qquad (6)$$

Table 2. Experimental results for 10 samples

S. No	Iris Image	Extracted Template 20*480	Reshaped Template 100*100	Encrypted Template 100*100	Decrypted Template 20*480
1					
2					
3					
4					
5					
6					
7					
8					
9					
10					

where p_1 and p_2 are two templates of size n*m, μ_1 and σ_1 are mean and standard deviation of p_1 and μ_2 and σ_2 are mean and standard deviation of p_2. The results of normalized correlation coefficients between original and encrypted and original and decrypted are shown in Fig. 5.

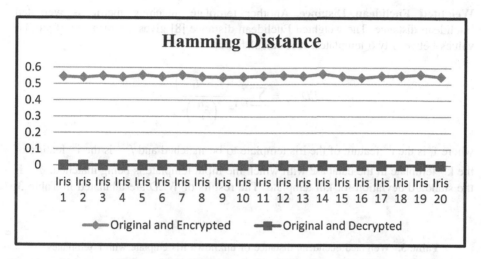

Fig. 4. Hamming distance between original and encrypted, original and decrypted iris templates

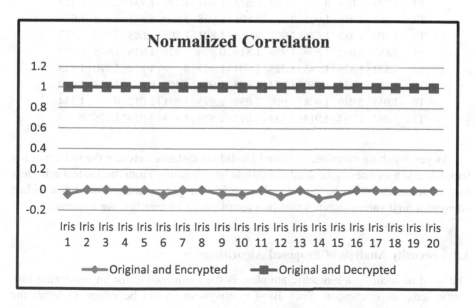

Fig. 5. Normalized correlation coefficient between original and encrypted, original and decrypted iris templates

As per the matching criterion required for template matching, a high normalized correlation is obtained for original and decrypted template pair while a low correlation value is obtained for original and encrypted pair. As shown in Fig. 5 the experimental values are encouraging and fulfill the design requirements of a good template encryption algorithm.

Weighted Euclidean Distance. Another template matching metric is weighted Euclidean distance. The weighted Euclidean distance [8] gives the measure of similar values between two templates. It is calculated as

$$WED(k) = \sum_{i=1}^{N} \frac{\left(f_i - f_i^{(k)}\right)^{\wedge} 2}{\left(\delta_i^{(k)}\right)^{\wedge} 2} \tag{7}$$

where f_i is the ith feature of the iris template to be matched and $f_i^{(k)}$ is the ith feature of the kth template in the database with which the input template is being matched. $\delta_i^{(k)}$ is the standard deviation for kth template. The results of WED are tabulated in Table 3.

Table 3. Weighted euclidean distance of unknown iris emplate with k templates

	T1	T2	T3	T4	T5	T6	T7	T8	T9	T10
T1	0	1.797	1.770	1.955	1.787	1.879	1.863	1.802	1.956	1.942
T2	1.810	0	1.811	1.961	1.885	2.008	1.834	1.978	2.053	1.797
T3	1.779	1.807	0	1.907	1.673	1.861	1.746	1.731	1.979	1.924
T4	1.966	1.958	1.907	0	1.898	1.958	1.930	1.813	1.910	1.946
T5	1.800	1.885	1.676	1.901	0	1.821	1.808	1.683	2.040	2.027
T6	1.887	2.002	1.858	1.956	1.815	0	1.926	1.916	1.975	2.074
T7	1.871	1.829	1.745	1.928	1.804	1.927	0	1.822	1.870	1.823
T8	1.860	1.859	1.750	1.774	1.895	1.921	2.108	0	1.931	1.909
T9	1.965	2.049	1.978	1.909	2.036	1.823	1.871	2.012	0	1.734
T10	1.942	1.784	1.914	1.935	2.013	1.819	1.814	1.939	1.726	0

As per matching criterion, weighted Euclidean distance between the unknown iris template and templates in the database should be minimum. From the Table 3 it is clear that value comes out to be zero if unknown iris template matches with kth template whereas a high values signifies that two templates to be matched are different.

4.2 Security Analysis of Proposed Algorithm

The need of securing a biometric template is the main motive for implementing biometric template encryption. Such Bio-Cryptosystem should be robust to fulfill the requirements of cryptographic security.

Cryptographic Security. Under cryptographic security the ability of proposed biometric template encryption algorithm to resist cryptanalysis is studied. Cryptanalysis is the study of methods to intrude into cryptographic security systems and gain access to the encrypted template even if the security keys are unknown. Cryptographic security for the proposed algorithm includes its ability to resist cryptanalysis attacks like key related attacks, statistical attacks, differential attacks [18].

Key space and key sensitivity analysis:

Key Space Analysis. To understand the key space analysis, Kirchhoff's principle is used. According to this principle, cryptanalyst has complete information regarding the algorithm and transmission channel except for the encryption and decryption key which are kept secret. If the keys are chosen poorly or the key space is too small, even a very well designed algorithm can be breached.

In the proposed algorithm, the key structure is $(n_1, a_1, b_1, x_1, a_2, b_2, x_2)$ for the encryption process where n_1 is the number of Arnold iterations for shuffling of biometric template and a_1, b_1, x_1 and a_2, b_2, x_2 are the initial conditions used for generation of pseudorandom matrix. For the decryption process and Key sensitivity analysis is done to find the difference between two encrypted outputs, when the same input image is encrypted using slightly different keys. Key sensitivity analysis also estimates the data loss when decryption is performed with a slightly different set of keys. Thereby, to estimate the key sensitivity encryption and decryption are performed using incorrect keys.

To analyze the first parameter of key sensitivity, the original encryption key E_K: (60, 1.5, 4.0, 0.3, 2.0, 3.5, 0.4) is slightly modified as E_{K1}: (60, 1.5, 4.0, 0.3, 2.0, 3.500000000000001, 0.4), E_{K2}: (60, 1.5, 4.0, 0.3, 2.0, 3.5, 0.4000000000000001), E_{K3}: (60, 1.5, 4.0, 0.3, 2.00000000000001, 3.5, 0.4), E_{K4}: (60, 1.5, 4.0, 0.30000000001, 2.0, 3.5, 0.4). Analysis has shown that initial conditions of logistic map (a, b) exhibit the key sensitivity of approximately 10^{-14} to 10^{-15}, while x exhibit a key sensitivity of 10^{-16}. Thereby, the total key space of proposed algorithm is around 10^{90}.

Key Sensitivity Analysis. Key sensitivity analysis is used to find the difference between the encrypted images when the same original image is encrypted using slightly different keys as shown in Fig. 6.

The above Fig. 6 demonstrates key sensitivity analysis at encryption end has been done. C_1 shows the encrypted template with original set of keys E_K while C_2, C_3, C_4, C_5 shows encrypted biometric templates with slightly modified set of keys $E_{K1}, E_{K2}, E_{K3}, E_{K4}$. Key sensitivity analysis difference between original key encrypted image and slightly modified key encrypted template is calculated in Table 4 to show that there is a significant change in templates with slight variation in the key structure.

Further correlation coefficient has been calculated between the encrypted templates tabulated in Table 4. A low value of correlation coefficient shows that templates encrypted with slightly different keys are looking similar but they are significantly different from each other which show that proposed algorithm is highly key sensitive.

Key sensitive analysis at decryption end signifies that when decryption is done with slightly incorrect keys there is no leakage of information about original image. Figure 7 shows the key sensitivity analysis at decryption end.

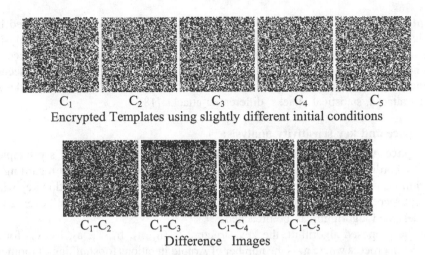

C₁　　　　C₂　　　　C₃　　　　C₄　　　　C₅

Encrypted Templates using slightly different initial conditions

C₁-C₂　　　C₁-C₃　　　C₁-C₄　　　C₁-C₅

Difference　Images

Fig. 6. Key sensitivity analysis at encryption end

Table 4. Correlation values for encrypted templates with original and slight different set of keys.

	Correlation Between C1 and			
	C2	C3	C4	C5
Template 1	−0.3485	−0.4961	−0.2538	−0.2759
Template 2	−0.4906	−0.4961	−0.2535	−0.2758
Template 3	−0.3488	−0.4964	−0.2540	−0.2762
Template 4	−0.3488	−0.4947	−0.2523	−0.2750
Template 5	−0.3482	−0.4959	−0.2535	−0.2758
Template 6	−0.3488	−0.4964	−0.2540	−0.2762
Template 7	−0.3489	−0.4964	−0.2541	−0.2763
Template 8	−0.3487	−0.4964	−0.2539	−0.2761
Template 9	−0.3488	−0.4964	−0.2540	−0.2762
Template 10	−0.3485	−0.4962	−0.2537	−0.2760
Template 11	−0.3488	−0.4962	−0.2538	−0.2760
Template 12	−0.3481	−0.4956	−0.2537	−0.2757
Template 13	−0.3487	−0.4964	−0.2539	−0.2762
Template 14	−0.3488	−0.4964	−0.2540	−0.2762
Template 15	−0.3488	−0.4964	−0.2541	−0.2762
Template 16	−0.3485	−0.4962	−0.2538	−0.2759
Template 17	−0.3487	−0.4964	−0.2539	−0.2761
Template 18	−0.3486	−0.4962	−0.2538	−0.2760
Template 19	−0.3477	−0.4954	−0.2530	−0.2750
Template 20	−0.3488	−0.4964	−0.2540	−0.2762

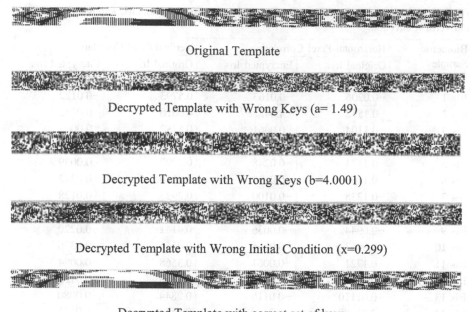

Original Template

Decrypted Template with Wrong Keys (a= 1.49)

Decrypted Template with Wrong Keys (b=4.0001)

Decrypted Template with Wrong Initial Condition (x=0.299)

Decrypted Template with correct set of keys

Fig. 7. Key sensitivity analysis at decryption end

Key sensitivity analysis at decryption end is shown in Fig. 7 where slight modification is done in any key of key structure yields a distorted output which means that proposed algorithm is highly key sensitive in nature at decryption end.

Statistical Attack Analysis. The resistance of proposed biometric template encryption algorithm against the statistical attacks can be determined by predicting the relationship between original biometric template and the encrypted biometric template. Correlation coefficient analysis has been done between original and encrypted biometric template pair.

To find the relationship between the original and encrypted biometric template, correlation coefficient is calculated in Table 5 among adjacent pixels in horizontal and vertical direction. A high correlation value is obtained for original biometric images in comparison to relatively low value for encrypted images. However in the proposed algorithm instead of biometric images biometric template has been used.

As tabulated in Table 5 the value of horizontal and vertical correlation obtained for encrypted biometric template is significantly low as compared to the original biometric template. As seen in the Table 5 correlation values for original template is also low owing to the fact that the biometric template is combination of 0's and 1's. Significant low value in encrypted format signifies that the correlation among adjacent pixels has further reduced indicating that there is no significant relationship between original and encrypted template pair.

Table 5. Correlation coefficient value of original and encrypted biometric template in horizontal and vertical direction.

Biometric Sample	Horizontal Pixel Correlation		Vertical Pixel Correlation	
	Original Iris Template	Encrypted Iris Template	Original Iris Template	Encrypted Iris Template
Iris 1	−0.0206	−0.0163	0.5793	−0.0122
Iris 2	−0.1552	−4.6422e^{-04}	0.6016	−0.0212
Iris 3	−0.1104	−0.0061	0.5182	0.0043
Iris 4	−0.0644	−3.0742e^{-04}	0.4669	−0.0243
Iris 5	−0.1174	−0.0248	0.3497	0.0039
Iris 6	−0.1108	0.0092	0.4140	−0.0312
Iris 7	−0.1238	−0.0100	0.2923	0.0128
Iris 8	−0.1196	−0.0096	0.4065	0.0184
Iris 9	−0.0944	−0.0036	0.4141	−0.0220
Iris 10	−0.0463	−0.0105	0.4165	−0.0376
Iris 11	−0.1721	−0.0061	0.3568	−0.0084
Iris 12	−0.1276	−0.0011	0.4768	−0.0135
Iris 13	−0.01110	−0.0116	0.2844	−0.0080
Iris 14	−0.01483	−0.0119	0.4484	0.0171
Iris 15	−0.1171	−0.0184	0.3991	0.0040
Iris 16	−0.1113	−0.0195	0.4088	2.2148e^{-04}
Iris 17	−0.1176	−0.0222	0.3524	−0.0027
Iris 18	−0.1264	0.0080	0.3964	−0.0088
Iris 19	−0.1156	−0.0059	0.2904	−0.0124
Iris 20	−0.1315	0.0015	0.3509	−0.0037

Differential Attack Analysis. For the proposed algorithm, the ability to withstand differential attacks is measured on the basis of two parameters Net Pixel Change Rate (NPCR) and Unified Average Changed Intensity (UACI) shown in Fig. 8. For calculating NPCR and UACI any one pixel value in the template is changed and then the original template and modified template are encrypted using the same encryption algorithm to measure the amount of deviation between encrypted templates when original templates differ slightly.

Mathematically, to calculate NPCR and UACI following things are considered. C^1 denotes the original template and C^2 denotes the modified template

$$D(i,j) = \left\{ \begin{array}{ll} 0 & if C^1(i,j) = C^2(i,j) \\ 1 & if C^1(i,j) \neq C^2(i,j) \end{array} \right\} \tag{8}$$

$$NPCR = \sum_{i,j} \frac{D(i,j)}{N*N} * 100\,\% \tag{9}$$

$$UACI = \frac{1}{N*N} \sum_{i,j} \frac{|C^1(i,j) - C^2(i,j)|}{255} * 100\,\% \tag{10}$$

where i, j denotes the pixel position, N* N denotes the size of C^1 and C^2.

Figure 8 shows the NPCR and UACI values calculated for the proposed algorithm. A very high value of NPCR and low value of UACI depicts that there is no relationship between two encrypted templates indicating efficient encryption of the biometric templates.

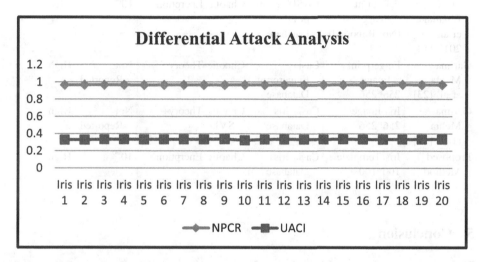

Fig. 8. NPCR and UACI

State of Art Comparison with Other Similar Existing Methods. Some similar work has been reported that uses biometric images for encryption. A comparative table is presented in Table 6 for biometric image encryption and biometric template encryption. The comparative Table 6 indicates that the proposed method has competitive performance in terms of data to be encrypted that whether image or template, size of the data to be encrypted and key space analysis.

The algorithms were implemented in MATLAB R2011b (Math Works) software on a CPU@ 2.3 GHz, 4 Gb RAM, 64 bit operating system. Encouraging computational time of 0.6 s is achieved using this method of template encryption without compromising other parameters like key sensitivity, key space. Although the size of the template to be encrypted is considerably small as compared to the image used in previous papers.

Table 6. Comparison Table for encryption

	Biometric Data used for encryption/Size	Source of the Data to be encrypted	Encryption method used	Key Space	Key sensitivity
Gaurav Bhatnagar et al. 2014 [22]	Fingerprint image 300*300	CASIA Fingerprint Database	Chaotic Encryption +Fractional Random Wavelet Transform	10^{102}	High
Gaurav Bhatnagar et al. 2012 [10]	Fingerprint image 300*300	CASIA Fingerprint Database	Fractional Wavelet Packet Transform	10^{84}	High
Gaurav Bhatnagar et al. 2012 [14]	Palmprint Image (Not Reported)	CASIA Palm print Database	Chaotic Encrption	10^{84}	High
Garima Mehta et al. [23]	Fingerprint Image 256*256	Casia Fingerprint Database	Chaotic Theory	Not Reported	High
Garima Mehta et al. [24]	Iris Image 256*256	Casia Iris Database	Chaotic Theory+ SVD	Not Reported	High
Proposed Method	Iris Template 100*100	Casia Iris Database	Chaotic Encrption	10^{90}	High

5 Conclusion

This paper proposes a security solution for secure transmission and secure storage of iris biometric template. Unlike conventional methods where the raw biometric images are encrypted, this paper proposes a method for encrypting biometric templates for secure storage and secure transmission over unsecured channels. Since the biometric template has much smaller size than the original biometric image, the computational complexity is reduced making this method suitable for real time applications. The experimental results are convincing and show that the proposed method has low computational complexity and also meets the requirements of template matching criteria. Template matching parameters like hamming distance, weighted Euclidean distance, and normalized correlation coefficient between original and encrypted template as well as original and decrypted gives the encouraging results. Further, the proposed algorithm also give good results against key related attacks using key space and key sensitivity analysis, against statistical attacks using correlation coefficient analysis and against differential attacks using NPCR and UACI. Some future scope of this work may be exploring the possibility the use of other biometric features for encryption and encryption methods in other domains.

References

1. Jain, A.K., Nandakumar, K., Nagar, A.: Biometric template security. EURASIP J. Adv. Sig. Process. **2008**, 17 (2008)
2. Daugman, J.: High confidence recognition of persons by test of statistical independence. IEEE Trans. PAMI **15**, 1148–1160 (1993)
3. Jain, Anil K., Ross, A., Parbhakar, S.: An Introduction to Biometrics Recognition. IEEE Trans. Circ. Syst. video Technol. **14**(1), 4–20 (2004)
4. Tian, J., Yang, X.: Biometric Recognition Theory and Application. Publishing House of Electronics Industry, Beijing (2005). ISBN 9787302184195
5. Prabhakar, S., Pankanti, S., Jain, A.K.: Biometric recognition: security and privacy concerns. Proc. IEEE Secur. Priv. **1**(2), 33–42 (2003)
6. Lian, S.: Multimedia Content Encryption: Techniques and Applications. CRC Press, Boca Raton (2008)
7. Huang, M.-Y., et al.: Image encryption method based on chaotic map. In: International Computer Symposium (ICS), pp. 154–158 (2010)
8. Masek, L., Kovesi, P.: Thesis on "Biometric Identification System Based on Iris Patterns," The School of Computer Science and Software Engineering, the University of Western Australia (2003)
9. Daugman, J.: How iris recognition works. In: Proceedings of International Conference on Image Processing, vol. 1, pp. I-33–I-36 (2002)
10. Bhatnagar, G., Jonathan Wu, Q.M.: Chaos-based security solution for fingerprint data during communication and transmission. IEEE Trans. Instrum. Meas. **61**(4), 876–887 (2012)
11. Liu, R.: Chaos-based fingerprint images encryption using symmetric cryptography. In: IEEE International Conference Fuzzy System Knowledge Discovery, pp. 2153–2156 (2012)
12. Jain, A.K., Bolle, R., Pankanti, S. (eds.): Biometrics: Personal Identification in Networked Society. Kluwer Academic Publishers, Boston (1999)
13. Dachselt, F., Schwarz, W.: Chaos and cryptography. IEEE Trans. Circ. Syst. I: Fundam. Theory Appl. **48**(2), 1498–1508 (2001)
14. Bhatnagar, G., Jonathan, Q.M.: A novel chaotic encryption framework for securing palm print data. Procedia Comput. Sci. **10**, 442–449 (2012)
15. Bhatnagar, G., Wu, Q.M.J.: Enhancing the transmission security of biometric data using chaotic encryption. Mutimedia Syst. **20**(2), 203–214 (2014)
16. Taneja, N., Raman, B., Gupta, I.: Combinational Domain Encryption for Still Visual Data. Multimdia Tools Appl. **59**(3), 775–793 (2012)
17. Khan, M.K., Zhang, J., Alghathbar, K.: Challenge-response-based biometric image scrambling for secure personal identification. Future Generation Comput. Syst. **27**(4), 411–418 (2011)
18. Taneja, N., Raman, B., Gupta, I.: Chaos based cryptosystem for still visual data. Multimedia Tools Appl. **16**(2), 281–298 (2012)
19. G Peterson Arnold's cat map (1997). Available from http://online.redwoods.cc.ca.us/instruct/darnold/maw/catmap3.html
20. Soutar, C., Roberge, D., Stoianov, A., Gilroy, R., Vijaya Kumar, B.V.K.: Biometric Encryption, ICSA Guide to Cryptography. McGraw Hill, New York (1999)
21. Huang, M.-Y., et. al.: Image Encryption Method Based on Chaotic Map. In: International Computer Symposium (ICS), pp. 154–158 (2010)
22. Bhatnagar, G., Jonathan-Wu, Q.M., Raman, B.: A new fractional random wavelet transform for fingerprint security. IEEE Trans. Syst. Man Cybern. Part A Syst. Hum. **42**(1), 262–275 (2012)

23. Mehta, G., Dutta, M.K., karasek, J., Kim, P.S.: An efficient and lossless fingerprint encryption algorithm using henon map and arnold transformation. In: ICCC 2013, pp. 485–489 (2013)
24. Mehta, G., Dutta, M.K., Kim, P.S.: An efficient and secure encryption scheme for biometric data using holmes map and singular value decomposition. In: 2014 International Conference on Medical Imaging, m-Health and Emerging Communication Systems (MedCom), pp. 211–215. IEEE (2014)
25. Menezes, A.J., Van Oorschot, P.C., Vanstone, S.A.: Handbook of Applied Cryptography. CRC Press, Boca Raton (1996)

Author Index

Printed in the United States
By Bookmasters